Smart Sensors, Measurement and Instrumentation

Volume 37

Series Editor

Subhas Chandra Mukhopadhyay, School of Engineering, Macquarie University, Sydney, NSW, Australia

More information about this series at http://www.springer.com/series/10617

Shantanu Pal

Internet of Things and Access Control

Sensing, Monitoring and Controlling Access in IoT-Enabled Healthcare Systems

Springer

Shantanu Pal
Department of Computing
Macquarie University
Sydney, NSW, Australia

ISSN 2194-8402 ISSN 2194-8410 (electronic)
Smart Sensors, Measurement and Instrumentation
ISBN 978-3-030-64997-5 ISBN 978-3-030-64998-2 (eBook)
https://doi.org/10.1007/978-3-030-64998-2

This Springer imprint is published by the registered company Springer Nature Switzerland AG
The registered company address is: Gewerbestrasse 11, 6330 Cham, Switzerland

Preface

The emergence of the Internet of Things (IoT) has already produced significant changes in our everyday lives, where everything and anything can be connected and communicated in the cyber-physical world. With the proliferation of smart mobile devices, intelligent sensors, wearable devices and ubiquitous Internet and cloud computing, the use of the IoT is growing at an increasing rate. However, this growth poses numerous challenges for the designers and users of these systems. One significant challenge is the provision of security within the IoT. The high mobility of *things*, the potential scale of the systems in the number of *things* and users combined with dynamic network topology and wireless communication mediums create a challenging environment. This is only exacerbated by the limitations in device memory, battery-life and processing capacity, arguing against the use of 'heavy-weight' security architectures.

The motivation behind this book is driven by the need to examine security mechanisms for large-scale IoT systems, in particular, the need for access control, identity management, the delegation of access rights and trust management within such systems. We focus on sensing, monitoring and controlling access in IoT-enabled healthcare systems. We intend to show how a policy-based approach provides fine-grained access for authorized users to services while protecting valuable resources from unauthorized access. We explain a hybrid approach by employing attributes, roles and capabilities for our authorization design. We show the use of attributes for role membership assignment and in permission evaluation. Membership of roles grants capabilities. The capabilities which are issued may be parameterized based on further attributes of the user and are then used to access specific services provided by IoT devices. This significantly reduces the number of policies required for specifying access control settings. We also explore the provision of an identity-less, asynchronous and decentralized delegation model for the IoT leveraging the advantages of blockchain technology. We consider the uncertainty that exists in IoT systems and introduce a trust model based on attribute-based identity. Notably, the significant contribution of this work is the use of attributes for identifying an entity rather than depending upon the unique concrete identity of that entity. That said, we use attributes to validate an entity rather than depending upon unique identities.

The work presented in this book is obtained from a doctoral research conducted at Macquarie University, Sydney, Australia. The motivation of this research was to design and development of an access control architecture for the large-scale IoT systems, in particular, keeping a view to the sensing, monitoring and controlling access in IoT-enabled healthcare systems. The author is obliged to *acknowledge* Associate Professor Michael Hitchens, Professor Vijay Varadharajan, Dr. Tahiry Rabehaja and Ambrose Hill for their help and support in this research work. The research was supported by the financial, academic and technical assistance of Macquarie University, Sydney, Australia.

Sydney, Australia Shantanu Pal

Contents

Chapter 1
Introduction

Abstract This chapter introduces the fundamental of the Internet of Things (IoT) and the need for access control in such systems. This, in turn, will help to explain the need for a scalable, dynamic, and flexible access control architecture for the IoT.

1.1 The Internet of Things

The Internet of Things (IoT) [1] is a paradigm shift where anything and everything in the physical and virtual worlds can be the part of the network. The term 'IoT' was popularized by the innovative work of the Massachusetts Institute of Technology (MIT) Auto-ID Centre. The first documented evidence of the use of the term *'The Internet of Things'* was by Kevin Ashton, the co-founder of the MIT Auto-ID Centre, in the year 1999 [2]. The IoT connects all the devices in a physical domain with the Internet to communicate with each other for faster and easier service. The IoT represents a view in which the traditional Internet extends into real-world objects (e.g. food, clothing, furniture, paper, landmarks, monuments, etc.) and enables each the ability to gather, process and act on information in a *smarter* way. These objects, acting as sensors or actuators, are able to interact with each other in order to reach a common goal (e.g. quality and service) by connecting all *smart things* to the current Internet. Therefore, the perspective for the IoT is to deploy a ubiquitous society where the users (i.e. people) and the various objects (i.e. everything that is addressable and communicable) will be connected over a network platform to leverage the benefits for both society and technology on a large scale, so that human users are unobtrusively assisted by technology in performing everyday activities [3].

IoT systems may deal with high volumes of data. This data can be particularly sensitive, as it may include health, location and other highly personal information. IoT systems are very large and dynamic in nature, and offer services that are related to human users or other *things* that constitute such systems. We envisage a growth of the IoT where it encompasses a significant range of human and social activities, e.g. commerce, leisure, healthcare and transport. Activities that are currently not digitally enabled will be supported and others expanded by the edge intelligence and ubiquity of the devices that constitute the IoT. For example, shopping may be

S. Pal, *Internet of Things and Access Control*, Smart Sensors, Measurement and Instrumentation 37, https://doi.org/10.1007/978-3-030-64998-2_1

enhanced by services offered by *things* deployed by the retailers, contacting user devices and offering information and discounts. Current services, e.g. e-tickets, may be enhanced by sensors detecting e-ticket holders and controlling physical access on that basis. Healthcare may be expanded by a range of sensors attached to a person [4].

Situations e.g. those just described, and others left to the reader's imagination, will require an even greater number of *things* than are currently deployed, and users devices that are likewise a part of the IoT. Devices and *things* may, over their lifetime, interact with a vast range of other *things*. Such interactions may be fleeting and may only occur once between a particular pair of *things* or be much more frequent and long term. *Things* will likely be highly mobile, especially devices, moving from administrative domain to another administrative domain. These domains will have to establish policies and mechanisms to enable them to deal with devices and *things* about which they have very limited, if any, previous information. This poses significant challenges in securing IoT systems [5].

This vision of the IoT implies that knowing the identities of individual entities before an interaction is, in many cases, impractical. In a system like the IoT, where there may be a large number of devices, being able to easily identify them, both uniquely and as groups, is challenging. In an environment consisting of a multitude of users, each possessing a large range of smart devices, questions of identity and access are paramount. Smart devices will come in many forms and provide a vast range of services to both their users and other entities within the IoT. Users will wish to maintain their privacy, efficiently manage their devices and quickly and precisely obtain the services they require. The devices will, for the foreseeable future, present relatively low-power capacities and solutions must be tailored to this.

IoT implementations are commonly based on sensing and wireless communication technologies e.g. Radio Frequency Identification (RFID), Wireless Sensor Networks (WSNs), WiFi, AdHoc Peer-to-Peer (P2P), 3G, 4G, IEEE 802.15.x communication standards e.g. ZigBee, Wireless Local Area Network (WLAN) and Wireless Body Area Network (WBAN), etc. [6]. The basic communication protocols for the IoT are predicated on the low power consumption of the devices. For example, PowerLine Communication protocol (PLC), IPV6 Low Power Wireless Personal Area Network protocol (6LowPAN), Routing Protocol for Low power lossy networks (RPL), ZigBee Smart Energy 2.0, EIST M2M Architecture and Message Queue Telemetry Transport (MQTT) protocol [7, 8]. These technologies help the *things* in automatic identification of other *things* they are communicating with. Further, with the development of IPV6, it is convenient to assign unique digital identities and access to various digital information and services by the *things*.

The IoT has already made significant changes in our daily lives, with the enterprise, social, technological and individual benefits only expected to increase with the proliferation of IoT applications and services. At the same time the IoT poses new security and privacy challenges for users, devices, systems and *things* [9]. Protecting IoT systems is difficult due to the particular characteristics of IoT systems. These include extremely large numbers of users and devices, many of the latter possessing low computing power and limited storage capacity, the heterogeneous nature of these systems, the potentially transient nature of relationships and other factors [10].

In addition, the scale of the number of individual devices, applications, services, etc., and the lack of common standards and architectures increases the difficulty in employing traditional security approaches. Importantly, with the vast range of users and entities, the issues of controlling access to resources and even identifying the communicating entities, access control and identity management are crucial issues in providing secure and trustworthy IoT implementations. This is only heightened by the likelihood of interacting entities originating from multiple distinct jurisdictions.

1.2 Need for Access Control

Access control is one of the crucial aspects of security when considering the characteristics e.g. scale and heterogeneity in devices, users, applications and services, of an IoT system. It is used to control and regulate who (e.g. an entity) can view or use what (e.g. a resource). Access control helps to satisfy the security properties of confidentiality, integrity and availability [11]. In a large-scale and highly dynamic system like the IoT, how to protect sensitive information from unauthorized users and services is a significant issue. In other words, allowing a legitimate entity for accessing a resource in the IoT, is highly demanding.

With the emergence of the IoT, there has been a tremendous growth in the use of intelligent sensors and wearable devices. It is estimated that the number of devices connected to the Internet will be 28.5 billion in 2022, up from 18 billion in 2017. On an individual basis this will mean 3.6 networked connected devices per capita by 2022, compared to 2.3 in 2016. The average number of devices and connections per household and per Internet user is predicted to increase by 51% by 2022. This trend will also increase the annual global Internet traffic, which is predicted to reach 4.8 ZB (zetta-bit) per year by 2022 [12]. These devices will generate a high amount of data, including a user's personally identifiable information (PII) and confidential health information as well as the contextual information, e.g. location, date and time [13, 14].

Commonly, there is a tradeoff between placing the access control at the edge of the network, making use of the intelligence of the devices, and relying on centralized, but more easily managed, mechanisms [15]. On the one hand, low-powered devices, with limited memory capacity and restricted processing power are often unable to support implementation of traditional security mechanisms. On the other hand, centralization may have difficulty coping with the scale of the systems. From the communication point of view, heterogeneous network environments, wireless communication mediums, high mobility of *things*, dynamic network topology and availability of infrastructure for communication also pose significant challenges [16]. These limitations restrict where 'heavy-weight' security mechanisms can be applied directly into edge IoT devices. This situation is exacerbated by the use of resource-heavy protocols e.g. HTTP/HTTPS and TCP for communicating between devices. Different solutions are needed for the protocol stack for use with such constrained resources, further inhibiting the use of traditional security approaches. For example, the use of CoAP (Constrained Application Protocol) [17] or DTLS (Datagram Transport Layer Security) [18] security protocol over the 6LoWPAN based on IEEE 804.15.4 standard, could be an alternative.

As of today, there is no complete, coherent and fine-grained access control approach that can be deployed for an IoT system [19]. This presents significant challenges in developing a secure, robust and scalable IoT system with secure applications and services. Importantly, the present access control architectures are not prepared to fully integrate with the different layers of an IoT system and do not adequately cover its dynamic and autonomous communications characteristics. Traditional security and privacy challenges are mostly related to information leakage and potential data loss and control over services. However, they have now become more significant due to the range of threats and attacks [20]. Moreover, as noted above, the high level of heterogeneity in the IoT, combined with the variety of technologies, systems and applications, mobility, dynamic network topology and limited physical security of low power devices further introduces advanced security risks that are growing concerns and need to be addressed for this kind of systems in the future [21]. Hence, there is a significant need to revisit and rethink access control approaches for developing an IoT system in a structured and comprehensive way.

1.3 Motivation and Use Case Examples

In this section, we present simple use case example scenarios to show the motivation of our research.

In the current large-scale IoT systems it is difficult to manage and track who (e.g. user and device) is using the system and what (e.g. data and resource) they are connecting to and accessing. This limits a wider deployment of such systems and indicates the demand for developing a robust, scalable and secure IoT systems [22]. Now let us consider an environment that is composed of a vast number of IoT devices, users, applications and their associated services. For an example scenario, we select an IoT-enabled smart healthcare system—one of the major application domains for the IoT. According to the World Health Organization (WHO) Global Observatory for eHealth survey in 2015, there were 121 countries with national eHealth strategies [23]. This illustrates the requirement for integrated, systematic and electronic use of healthcare information by adopting enabling policies and mechanisms for both the patients and healthcare providers. Moreover, with the rapid expansion of the IoT, healthcare systems are expected to be easily accessible as well as remotely available [24].

An important element of this is the increased use of a wide range of wearable devices, including smart sensors, to enable automatic collection, storage and reporting of data and its use in diagnosis [25]. For instance, In 2013, 15% of the population in the United States owned a mobile phone-connected wearable device (e.g. Fitbit or smartwatch) [26, 27]. The number of applications compatible with wearable devices will only grow, with nearly 165,000 healthcare apps. available at present [28]. In

Australia people can store their health records e.g. allergies, past and current conditions and treatments, medicine details, pathology reports or diagnostic imaging scan reports using the MyHealthRecord [29] online service. These reports can be seen by the appropriate doctors, specialists and hospital staff online from anywhere at any time, for example, in an accident or emergency condition or even in a regulation medical visit. Similarly, through the National Health Portal (NHP) [30] in India, an initiative has been taken where citizens' health records will be integrated in a common network/grid for efficient monitoring of health entitlements by both public and private healthcare providers. According to the European Commission, up to 50% of European adults depend on online health information systems for searching and managing their health information [31].

Now consider in such an IoT-enabled healthcare facility where patient monitoring and possibly even treatment delivery is administered via smart *things*. Various wearable devices and sensors will be attached to the patients and accessed by healthcare professionals (and possibly others). The sensors and devices may be allocated to a patient on admission or at any stage during their stay at the facility. When they (i.e. the *things*) are allocated to a patient they will be registered in the system, including noting which patient they are assigned to. Access to the sensors will depend upon the policies of the facility and this may include different users having different levels of access to the same device (for example, a nurse may have read only access to a drug delivery device whereas a doctor may be able to alter the dosage).

The issue is how to identify all the entities in the system, provide a different level of access to different actors to the patient's medical sensors and to protect the patient's privacy. In a real-life hospital environment there may be hundreds of doctors, nurses, possibly thousands of patients and millions of sensors. The number of possible relationships in accessing a patient's private information is considerable. Now imagine three simple hypothetical scenarios as follows. Doctor A and Doctor B both work in a hospital H, as does Nurse C. Doctor A is a cardiac specialist caring for patient Alice and Doctor B is a cardiac specialist caring for patient Bob. Nurse C is appointed in the ward Y, where Alice is admitted. Bob is assigned to a different ward, X. Doctor D is another cardiac specialist, who is brought in to assist in an operation on Alice. From the access control perspective some of the concerns are:

- To allow a doctor (e.g. a cardiac specialist) to access only their (particular) patient's *things*. We do not want a specialist to gain access to every patient's medical *things*. For example, cardiac specialist Doctor A for patient Alice is allowed to access the cardiac-related medical *things* for Alice. Doctor A should not be allowed to access the same devices worn by Bob and may not even be allowed to access all devices attached to Alice (for example, a location tracker). While the desired situation could be achieved by authoring policies for every doctor-patient combination, this is not scalable and also requires a-priori knowledge of all the system users.
- Similarly, a nurse who is assigned to a particular ward, is able to access the clinical and related medical *things* of patients who are admitted in that particular ward. For example, nurse C in ward Y can access some of patient Alice's *things*. However, nurse C is not able to access any of Bob's *things*, as Bob is admitted in ward

X. Likewise, other appropriate policies and access control can be assigned for a patient's family and friends and other staff of the hospital. As patients move between wards of the hospital we do not wish to constantly revise the policies to reflect this.

- Being able to rapidly grant Doctor D access to the devices attached to Alice. In the event that an emergency arises and prompt action needs to be taken, D needs to access the information and functionality of the *things*. There may be insufficient time to enter D's identity in the system. A solution where *attribute-based* credentials could suffice for access would obviate the need for a-priori identity knowledge.

Thus, for the patients, healthcare providers and associated organizations it is a significant need to keep their patients' information confidential and secure the access of the devices while managing the scale of devices, identities and the nature of relationships in the IoT. An unauthorized access to these wearable devices (and connected medical equipment) can breach a patient's privacy and generate potential life-threatening attacks [32]. For example, an attack called MedJack (Medical Device Hijack) [33] allowed attackers to inject malware into medical devices and perform unauthorized activities, for instance, remotely stopping an x-ray machine working. It can be seen that appropriate access control and identity mechanisms are crucial in limiting the unauthorized actions and operations that an attacker can perform.

1.4 The Aim of the Book

The aim of the book is to examine the significant security challenges of access control and, in particular, the specific needs of access control for the IoT systems. There have been a number of proposals that discuss access control models and mechanisms for use in IoT systems [34–36]. These include well-known access control solutions e.g. Role-Based Access Control (RBAC) [37], Attribute-Based Access Control (ABAC) [38] and Capability-Based Access Control (CapBAC) [39], just to name a few. However, each of these mechanisms has its own advantages and disadvantages when applied to the IoT. For instance, RBAC provides fine-grained access control over the resources using explicit user-to-role mappings, however, RBAC itself is highly centralized and requires the definition of each user-to-permission relation for each resource that a user is to be allowed to access. This is challenging in a large and complex system like the IoT [35]. The employment of ABAC improves policy management by using attributes (e.g. name, age, location, etc.) rather than concrete identity. This is more flexible for the IoT as policies can be written based on the context (e.g. current time or a location). However, ABAC by itself provides no mechanism for controlling the number of policies required, e.g. by grouping together policies with the same attribute requirements. This is an issue in highly scalable systems e.g. the IoT [34].

CapBAC provides flexible access control. Users are provided with capabilities which identify the resources and operations on that resource that a user is allowed to access. This allows fine-grained access control. However, many of the existing (non-IoT) CapBAC mechanisms are centralized when validating access rights of subjects. Distributed CapBAC models e.g. [40, 41] have been proposed for use in the IoT. In this model, validation is performed inside the resource-constrained IoT devices (or a local management capacity) without there being any contact with a centralized authority. This allows a distributed approach, taking advantage of edge-intelligence, i.e. in-line with the nature of the IoT. However, these systems do not address the problems of managing the number of capabilities that will be required in a realistic IoT system, let alone the policy base needed to control their creation and distribution. The policy base that will need to be defined is likely to be large and dynamic. The scale and diverse nature of the IoT makes it difficult to specify, centrally and in advance, a complete set of access control policies.

Furthermore, most of the proposals are concerned with how identities can uniquely identify a particular entity. In other words, entities are defined by the unique identities. We argue that such an approach is not sufficiently flexible and fine-grained for a large and highly dynamic system like the IoT. Further, when considering issues e.g. policy management and delegation of access rights in the IoT, we need to be able to flexibly handle questions of identity. It cannot always be known in advance which users will access which services or devices or which devices will be available at the time when access is requested.

In addition, the nature of the IoT requires a fine-grained approach to access control, including in the handling of delegation. This means, just as with access control itself, the delegation of access rights needs to be governed by policies. The scale and nature of the IoT means that commonly used mechanisms by centralizing the control of delegation is likely to be impractical. With the exponentially growing number of IoT services, applications and devices, a fundamental issue is to ensure that only the entities that possess the appropriate rights are able to access resources. The IoT requires a flexible and fine-grained delegation model.

Returning to the use case scenario that we discussed in the previous section (i.e. Sect. 1.3), which is likely to be common in an IoT-enabled healthcare setting. A number of medical sensors (e.g. to monitor blood pressure, body temperature, etc.) are attached to a patient. The patient's doctors should be given access to the sensors to allow readings to be taken. Defining policies which give access to each doctor for each sensor will be time-consuming and hard to manage. Given the dynamic nature of both the IoT and healthcare situations, the set of sensors is likely to change in unpredictable ways, making managing their access particularly challenging. The problem of policy management, and particularly the number of policies that must be authored, requires addressing. The presented work in this book explains the design and development of an access control architecture for an IoT system that is capable of handling security using a minimum number of policies and dynamic identity management. Further, it shows how to achieve such a fine-grained access control design leveraging on the distributed nature of an IoT system.

1.5 Research Contributions

The major contribution of the research is the development of a novel access control architecture for the IoT that implies the development of secure access control, efficient identity management, and flexible access rights delegation. The contributions are listed below in short, and will appear in details in the following chapters of this book in their chronological order.

- *The design of a policy-based, fine-grained and partially decentralized access control architecture for the IoT.*
  In the IoT, security is a significant concern, with access control being one of the major issues. Towards this, we propose the design of a policy-based, fine-grained and partially decentralized access control architecture that allows fine-grained access for authorized users to services while protecting valuable resources from unauthorized access. In the IoT, with its open technologies and resource constrained nature of these devices (e.g. limited battery power, memory capacity and computational speed, etc.), managing the resources and users of the system and enforcing appropriate policies are difficult and challenging issues. The scale and diverse nature of the IoT makes it difficult to specify, centrally and in advance, a complete set of access control policies based on traditional access control mechanisms e.g. RBAC, ABAC, etc. To address this issue, we design a hybrid access control model employing *attributes*, roles and capabilities. We show that the proposed model can significantly reduce the required number of policies for granting access to a service (or resource) in an IoT system. In the design, the attributes are used for authentication and authorization of a legitimate user and services, rather than depending upon a concrete identity of an entity. In other words, in our model, the identity of an entity does not depend upon a unique concrete identity. We use attributes to parameterize capabilities for accessing specific services provided by IoT *things*. We apply attributes for role-membership assignment and in permission evaluation. Membership of roles grants capabilities. The capabilities which are issued may be parameterized based on further attributes of the user and are then used to access specific services provided by IoT *things*. We discuss the practicability of the proposed architecture with both symmetric and asymmetric key based approaches.
- *The development, implementation and evaluation of the proposed access control architecture.*
  Recent proposals for IoT access control do not provide any implementation information nor do they discuss the authorization process in detail. In this book, we provide a detailed discussion of the development of the proposed access control architecture using both symmetric and asymmetric key based approaches. We provide a detailed performance analysis of the employed symmetric key based approach in comparison with the asymmetric key based approach. We demonstrate that the proposed architecture could easily work with either approaches. We intend to examine the employment of the light-weight network authentication

protocol for constrained IoT devices which can be an alternative than enforcing heavy-weight security protocols for the IoT.

- *The examination of the notion of identity in the context of the IoT, which in turn helps to provide deeper insights into the different types of authentication and authorization issues that can be used for IoT access control.*
  In the IoT, it is difficult to predict, in advance, which entities will interact and require access to services and to precisely identify the exact services to which they will seek access. Therefore, we argue that depending upon a concrete identity of an entity in the IoT is not an ideal choice. Towards this, we address important questions concerning the nature of identity and identity management for such IoT systems. In the state of the art, there exist many approaches that discuss identity and its management, however examination of identity in the context of the IoT is still in its infancy. We introduce a formal model to represent IoT identity from a 'things-centric' approach. Importantly, we employ attributes for the authorization and authentication of an entity. We demonstrate that the use of attributes could be an alternative to represent IoT identity without depending upon the concrete identities of the entities.
- *The development of a dynamic and flexible delegation model to transfer access rights in an IoT scenario.*
  In IoT, access right delegation is one of the significant issues when addressing security. Access rights, in specific, governs who or what can view or use resources by the allocation of rights specified by certain policy enforcement. In an IoT system, it cannot always be possible to record in advance which users will wish to access which services or resources, or which devices will be available at the time access is requested. Entities still need some basis on which to determine whether to interact, including the bestowal and acknowledgement of access rights. To date, most models for access right delegation in IoT systems are built on the commonly used access control mechanisms e.g. RBAC and ABAC, which are not suitable to provide flexible and dynamic access control for IoT systems. We examine the need for an identity-less, asynchronous and decentralized delegation model for flexible and easy transfer of access rights in the IoT. We demonstrate that blockchain technology can be used to facilitate delegation of access rights to IoT systems governed by the generated capabilities issued by the smart contracts without the involvement of any trusted third party authentication. Significantly, we employ attributes to validate an entity in such delegation rather than depending upon the unique identity of the entities.
- *The foundation and examination of the notion of the trust in the context of the IoT access control mechanisms and propose a trust model supporting attribute-based identity.*
  Trust is an important aspect for establishing communication between different entities in uncertain conditions. Given the dynamic characteristics in the IoT, it is important to include mechanisms that can help in interactions between the *things* by overcoming this uncertainty. However, there are several challenges that need to be overcome including the resource constrained nature of the devices. In an IoT system billions and potentially trillions of devices will be interconnected with one

another which make the system more challenging for developing a flexible and secure trust model for the IoT systems. We propose a trust management model to reduce the uncertainties based on the past interactions. This will help to reinforce the confidence in trust value evaluation for the IoT. Our model employs subjective logic for modeling and evaluation of trust. Subjective logic is used to examine the use of direct experiences and recommendations to evaluate the final trust value for an entity in an IoT context. Importantly, we use attributes for representing an entity rather than depending upon their concrete unique identity in the trust management system. In other words, an entity is evaluated based on the attributes that they posses. We demonstrate the practicability of the proposed trust model with IoT-based real-world scenarios.

The rest of the book is organized as follows. In Chap. 2, we discuss the state of the art research background and related work in IoT and its access control. In Chap. 3, we discuss the development of an access control architecture for the IoT. We propose a fine-grained and flexible access control architecture for large-scale systems like the IoT based on the interactions between the entities and service discovery. In Chap. 4, we explore the detailed technical implementation along with the practical test-bed used for the experiments and illustrate the numerical evaluation of the proposed access control architecture. In Chap. 5, we discuss the notion of identity for the IoT. We examine the feasibility to incorporate such an identity model to achieve both fine-grained and flexible system design for large-scale IoT systems. In Chap. 6, we discuss the importance of access rights delegation in the context of the IoT. We demonstrate the feasibility of our model through use-case examples and analyze the performance with a physical proof of concept testbed implementation using blockchain technology. In Chap. 7, we examine how the notion of trust can be used for access control in the IoT. We explicitly take into consideration the uncertainty that exists in an IoT system. Finally, in Chap. 8, we conclude the book by summarizing our research findings.

References

1. N. Gershenfeld, R. Krikorian, D. Cohen, The internet of things. Sci. Amer. **291**(4), 76–81 (2004)
2. K. Ashton, That 'Internet of Things' thing. RFID (2009), http://www.rfidjournal.com/articles/view?4986
3. S. Li, L. Xu, S. Zhao, The internet of things: a survey. Inf. Syst. Front. **17**(2), 243–259 (2015). http://dx.doi.org/10.1007/s10796-014-9492-7
4. W. Wang, J. Li, L. Wang, W. Zhao, The internet of things for resident health information service platform research, in *IET International Conference on Communication Technology and Application (ICCTA)* (IET, 2011), pp. 631–635. http://dx.doi.org/10.1049/cp.2011.0745
5. J. Lin, W. Yu, N. Zhang, X. Yang, H. Zhang, W. Zhao, A survey on internet of things: architecture, enabling technologies, security and privacy, and applications. IEEE Int. Things J. 1 (2017). http://dx.doi.org/10.1109/jiot.2017.2683200

6. A. Al-Fuqaha, M. Guizani, M. Mohammadi, M. Aledhari, M. Ayyash, Internet of things: a survey on enabling technologies, protocols, and applications. IEEE Commun. Surv. & Tutor. **17**(4), 2347–2376 (2015). http://dx.doi.org/10.1109/comst.2015.2444095
7. P. Misra, Y. Simmhan, J. Warrior, Towards a practical architecture for the next generation internet of things (2016). arXiv:1502.00797
8. J. Tan, S.G.M. Koo, A survey of technologies in internet of things, in *2014 IEEE International Conference on Distributed Computing in Sensor Systems* (IEEE, 2014), pp. 269–274. http://dx.doi.org/10.1109/dcoss.2014.45
9. M. Noura, M. Atiquzzaman, M. Gaedke, Interoperability in internet of things: taxonomies and open challenges. Mobile Netw. Appl. **24**(3), 796–809 (2019). https://doi.org/10.1007/s11036-018-1089-9
10. Y. Yang, L. Wu, G. Yin, L. Li, H. Zhao, A survey on security and privacy issues in internet-of-things. IEEE Int. Things J. 1 (2017). http://dx.doi.org/10.1109/jiot.2017.2694844
11. N. Fotiou, T. Kotsonis, G.F. Marias, G.C. Polyzos, Access control for the internet of things, in *2016 International Workshop on Secure Internet of Things (SIoT)* (IEEE, 2016), pp. 29–38. http://dx.doi.org/10.1109/siot.2016.010
12. CISCO, The Zettabyte Era: Trends and Analysis, https://www.cisco.com/c/en/us/solutions/collateral/service-provider/visual-networking-index-vni/vni-hyperconnectivity-wp.html. Accessed 1 Nov 2017
13. A. Nordrum, The internet of fewer things [News]. IEEE Spectrum **53**(10), 12–13 (2016). http://dx.doi.org/10.1109/mspec.2016.7572524
14. I. Yaqoob, E. Ahmed, I.A. Hashem, A.I. Ahmed, A. Gani, M. Imran, M. Guizani, Internet of things architecture: recent advances, taxonomy, requirements, and open challenges. IEEE Wireless Commun. **24**(3), 10–16 (2017). http://dx.doi.org/10.1109/mwc.2017.1600421
15. C. Medaglia, A. Serbanati, An overview of privacy and security issues in the internet of things, in *The Internet of Things*, ed. by D. Giusto, A. Iera, G. Morabito, L. Atzori (Springer, New York, 2010), pp. 389–395. http://dx.doi.org/10.1007/978-1-4419-1674-7_38
16. S. Sicari, A. Rizzardi, L.A. Grieco, A. Coen-Porisini, Security, privacy and trust in Internet of Things: the road ahead. Comput. Netw. **76**, 146–164 (2015). http://dx.doi.org/10.1016/j.comnet.2014.11.008
17. P.P. Pereira, J. Eliasson, J. Delsing, An authentication and access control framework for CoAP-based Internet of Things, in *IECON 2014 - 40th Annual Conference of the IEEE Industrial Electronics Society* (IEEE, 2014), pp. 5293–5299. http://dx.doi.org/10.1109/iecon.2014.7049308
18. T. Kothmayr, C. Schmitt, W. Hu, M. Brunig, G. Carle, A DTLS based end-to-end security architecture for the Internet of Things with two-way authentication, in *37th Annual IEEE Conference on Local Computer Networks – Workshops* (IEEE, 2012), pp. 956–963. http://dx.doi.org/10.1109/lcnw.2012.6424088
19. N. Zhang, S. Demetriou, X. Mi, W. Diao, K. Yuan, P. Zong, F. Qian, X. Wang, K. Chen, Y. Tian, C.A. Gunter, K. Zhang, P. Tague, Y.-H. Lin, Understanding IoT security through the data crystal ball: where we are now and where we are going to be (2017). arXiv:1703.09809.pdf
20. R. Khan, S.U. Khan, R. Zaheer, S. Khan, Future internet: the internet of things architecture, possible applications and key challenges, in *2012 10th International Conference on Frontiers of Information Technology (FIT)* (IEEE, 2012), 257–260. http://dx.doi.org/10.1109/fit.2012.53
21. L. Malina, J. Hajny, R. Fujdiak, J. Hosek, On perspective of security and privacy-preserving solutions in the internet of things. Comput. Netw. **102**, 83–95 (2016). http://dx.doi.org/10.1016/j.comnet.2016.03.011
22. Suhardi, A. Ramadhan, A survey of security aspects for internet of things in healthcare, in *Information Science and Applications (ICISA)*, ed. by K.J. Kim, N. Joukov. Lecture Notes in Electrical Engineering, vol. 376 (Springer, Singapore, 2016), pp. 1237–1247. http://dx.doi.org/10.1007/978-981-10-0557-2_117
23. W.H. Organization, eHealth at WHO (2015), http://www.who.int/ehealth/about/en/. Accessed 29 May 2017

24. Y. Bhatt, C. Bhatt, Internet of things in HealthCare, in *Internet of Things and Big Data Technologies for Next Generation Healthcare*, ed. by C. Bhatt, N. Dey, A.S. Ashour. Studies in Big Data, vol. 23 (Springer International Publishing, Berlin, 2017), pp. 13–33. http://dx.doi.org/10.1007/978-3-319-49736-5_2
25. G. Mujica, J. Portilla, T. Riesgo, Deployment strategies of wireless sensor networks for IoT: challenges, trends, and solutions based on novel tools and HW/SW platforms, in *Components and Services for IoT Platforms*, ed. by G. Keramidas, N. Voros, M. Hübner (Springer International Publishing, Berlin, 2017), pp. 133–154. http://dx.doi.org/10.1007/978-3-319-42304-3_8
26. Fitbit, https://www.fitbit.com/au/home. Accessed 6 July 2017
27. P. Krebs, D.T. Duncan, Health app use among us mobile phone owners: a national survey. JMIR mHealth uHealth **3**(4) (2015)
28. M. Maheu, V. Nicolucci, M. Pulier, K. Wall, T. Frye, E. Hudlicka, The interactive mobile app review toolkit (IMART): a clinical practice-oriented system (2017), pp. 1–13. http://dx.doi.org/10.1007/s41347-016-0005-z
29. A.D.H. Agency, My Health Record (2017), https://myhealthrecord.gov.au/. Accessed 29 May 2017
30. G. of India, National Health Portal (2017), https://www.nhp.gov.in//NHPfiles/national_health_policy_2017.pdf. Accessed 10 June 2017
31. E. Commission, eHealth Action Plan 2012-2020 (2012), https://ec.europa.eu/digital-single-market/en/news/ehealth-action-plan-2012-2020-innovative-healthcare-21st-century. Accessed 09 June 2017
32. K. Saleem, Z. Tan, W. Buchanan, Security for cyber-physical systems in healthcare, in *Health 4.0: How Virtualization and Big Data are Revolutionizing Healthcare*, ed. by C. Thuemmler, C. Bai (Springer International Publishing, Berlin, 2017), pp. 233–251. http://dx.doi.org/10.1007/978-3-319-47617-9_12
33. Wikipedia, Medical device hijack (2017), https://en.wikipedia.org/wiki/Medical_device_hijack. Accessed 02 June 2017
34. A. Ouaddah, H. Mousannif, A. Abou Elkalam, A. Ait Ouahman, Access control in the internet of things: big challenges and new opportunities. Comput. Netw. **112**, 237–262 (2017). http://dx.doi.org/10.1016/j.comnet.2016.11.007
35. Y. Zhang, X. Wu, Access control in internet of things: a survey (2016). arXiv:1610.01065.pdf
36. P. Samarati, S. de Vimercati, Access control: policies, models, and mechanisms, in *Foundations of Security Analysis and Design*, ed. by R. Focardi, R. Gorrieri. Lecture Notes in Computer Science (Chap. 3), vol. 2171 (Springer, Berlin, 2001), pp. 137–196. http://dx.doi.org/10.1007/3-540-45608-2_3
37. D.F. Ferraiolo, R. Sandhu, S. Gavrila, D.R. Kuhn, R. Chandramouli, Proposed NIST standard for role-based access control. ACM Trans. Inf. Syst. Secur. **4**(3), 224–274 (2001). http://dx.doi.org/10.1145/501978.501980
38. E. Yuan, J. Tong, Attributed based access control (ABAC) for web services, in *Proceedings of the IEEE International Conference on Web Services*, ser. ICWS '05 (IEEE Computer Society, Washington, DC, USA, 2005), pp. 561–569. http://dx.doi.org/10.1109/icws.2005.25
39. S. Gusmeroli, S. Piccione, D. Rotondi, A capability-based security approach to manage access control in the internet of things. Math. Comput. Modell. **58**(5–6), 1189–1205 (2013). http://dx.doi.org/10.1016/j.mcm.2013.02.006
40. J.L. Hernández-Ramos, A.J. Jara, L. Marín, A.F. Skarmeta Gómez, DCapBAC: embedding authorization logic into smart things through ECC optimizations. Int. J. Comput. Math. **93**(2), 345–366 (2016). http://dx.doi.org/10.1080/00207160.2014.915316
41. J. Hernandez-Ramos, A. Jara, L. Marın, A. Skarmeta, Distributed capability-based access control for the internet of things. J. Int. Serv. Inf. Secur. **3**(3/4), 1–16 (2013), http://isyou.info/jisis/vol3/no34/jisis-2013-vol3-no34-01.pdf

Chapter 2
Background and Related Work

Abstract This chapter presents the background and related work in IoT and access control technology. We intend to present an overview of the state of the art IoT paradigm and, in particular, an examination of the available access control solutions in the existing literature. We also provide insight on IoT security, identity, access right delegation and the notion of trust when building an IoT access control architecture.

2.1 The IoT Paradigm

There are several definitions of the IoT that have been presented. For instance, Atzori et al. [1] define *things* from three perspectives e.g. middleware service, sensors and information. Buyya et al. [2] present a user-oriented definition of the IoT regardless of communication protocols and IoT environments: *"interconnection of sensing and actuating devices providing the ability to share information across platforms through a unified framework, developing a common operating picture for enabling innovative applications. This is achieved by seamless ubiquitous sensing, data analytics and information representation with cloud computing as the unifying framework"*.

Compared to [1, 2], Tan and Wang [3] define the IoT from the viewpoint of communication, social, environment and user contexts, as follows: *"things have identities and virtual personalities operating in smart spaces using intelligent interfaces to connect and communicate within social, environment, and user contexts"*. Unlike the approaches of [1–3], Haller et al. [4] define the IoT independently of technology and platforms. This definition is derived from a mobility and service integration perspective: *"a world where physical objects are seamlessly integrated into the information network, and where the physical objects can become active participants in business processes. Services are available to interact with these 'smart objects' over the Internet, query their state and any information associated with them, taking into account security and privacy issues"*. Unlike others, Davoli et al. [5] discuss the IoT from a network point of view: *"the IoT can be defined as a 'network of networks' of physical devices connected in an Internet-like structure, thus enabling them to collect, exchange and process data"*. A list of other definitions for the IoT can be found in [6].

S. Pal, *Internet of Things and Access Control*, Smart Sensors, Measurement and Instrumentation 37, https://doi.org/10.1007/978-3-030-64998-2_2

In summary, the IoT is not just a cyber-physical system for measuring state information and doing automatic computation. It is more a networking infrastructure that combines the digital and physical worlds together. Therefore, when we address the security characteristics of IoT, we need to consider a wider aspect of scenarios combining architectures, users, communications, technologies and applications.

2.1.1 Architecture

Various IoT architectures have been proposed in the literature, for example [7–11]. Many of them proposed a three-layer architecture (e.g. [12–14]) composed of application, network and perception layers. A few of them (e.g. [15]) propose a four-layer architecture consists of sensing layer, network layer, service layer and application-interface layer. However, there is no generic architecture for the IoT that has converged to a commonly-used reference model. In contrast to the three and four layers architectures, [9] argues for the support for a five-layer architecture for IoT applications and services. The layers used there are, from bottom to top, objects, object abstractions, service management, application and business. CISCO provides a reference architecture for the IoT by enhancing the traditional three-level and five-level models which is composed of seven layers [16]. In this architecture, the layers from bottom to top are physical devices and controllers, connectivity, edge (fog) computing, data accumulation, data abstraction, applications and, collaboration and processes.

We argue that the functional components of an IoT architecture should encapsulate the diverse security requirements and various security issues of this context. The architecture should enable the achievement of security for devices, networks, data repository, services, applications and users. Note, in order to explore various IoT security issues and requirements for IoT security provisioning, throughout this book we use a four-layer reference architecture that of [15] (cf. Fig. 2.1). Next, we briefly describe each layer.

- **Sensing Layer:** The first layer is composed of smart IoT sensing devices e.g. smart phones, RFID tags, sensors and actuators, etc. These components are able to automatically sense, collect and measure the various physical parameters e.g. temperature, humidity, location etc. Devices can store collected information inside themselves and sensors can store information into predefined sensor hubs (e.g. a microcontroller unit) to process them. The major functionalities of this layer are data sensing and data acquisition. Standardized plug-and-play mechanisms can be used with the various sensing devices. Furthermore, considering the scale of the number of *things* in an IoT system, sensing devices may be deployed simultaneously or over time according to the environmental context and practical requirements [17].
- **Network Layer:** The second layer is the network layer. This layer is composed of different wired and wireless networks, cloud computing services and big data

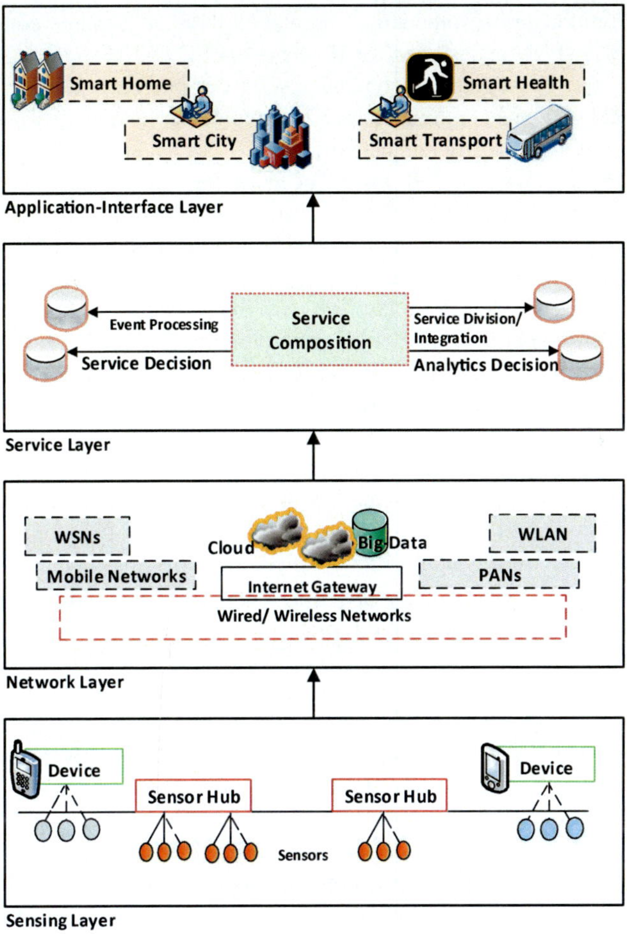

Fig. 2.1 The functional layers of a four-layer IoT architecture

repositories. Major functionalities of this layer include data aggregation, Quality of Service (QoS), scheduling, etc. It is also responsible for transmitting data to the next IoT architectural layer. The networks in this layer potentially combine heterogeneous equipment and help to transmit data among different components within this layer (and to the next architectural layer) using technology including 3G, 4G, GSM (Global System for Mobile Communication), UMTS (Universal Mobile Telecommunications System), WiFi, Bluetooth, etc. The presence of cloud computing services and big data repositories enable a variety of different technologies to perform seamlessly by deploying, managing and scheduling of various network services [18]. Other commonly used technologies in this layer are IPv6, 6LoWPAN (IPv6 over Low-Power Wireless Personal Area Networks), and RPL (IPv6 Routing Protocol for Low-Power and Lossy Networks). 6LoWPAN is a dedicated communication protocol that can fit well with the resource-constrained IoT devices. 6LowPAN is designed for IPv6 over IEEE 802.15.4 and it is connected to the Internet via a gateway (typically WiFi or Ethernet). Similarly to 6LoWPAN, RPL facilitates communication in a resource constrained environment and specifically within constrained networks, e.g. WSNs [19].

- **Service Layer:** This is the third layer. The major functions of this layer are analysis and processing of data that is collected from the network management layer. The service composition layer is built based on middle-ware technology that assists with information exchange for IoT applications among heterogeneous objects without any specific hardware and software requirements. It is intended to meet the needs of applications, application programming interfaces (APIs) and various service protocols [20]. The major functional component of this layer is the service composition unit, which is responsible for event processing, creating service divisions, service monitoring, service configuration and performing various decision analytics according to the specific policy requirements and contextual information.
- **Application-Interface Layer:** The fourth layer is the application-interface layer which provides smart IoT services to end users. The major functional components of this layer are various applications which could be classified as, for example, smart home, smart city, smart transport, smart commerce and smart health, etc. [20–22]. This layer is responsible for providing various services and at the same time determines a set of message passing protocols at the application level [23]. This layer is also responsible for data presentation, application maintenance, application access control and updating software and security patches for those applications. Standard interfaces using HTTP and HTTPS are widely deployed for this layer. However, more dedicated resource contained application level protocols e.g. CoAP (Constrained Application Protocol), Advanced Message Queuing Protocol (AMQP), eXtensible Messaging and Presence Protocol (XMPP), etc. are also used in this layer [24, 25]. This layer exports the system's functionalities from the service layer to the end users. It may also use standard Web services (both for service protocol and service composition) to distribute the activities and services.

2.1.2 Characteristics

In this section, we outline some fundamental characteristics of an IoT system as follows [9, 26, 27]:

- **Resources Limitation:** IoT devices are, in general, more resource constrained in nature than traditional networked devices. These devices have limited battery power, memory capacity and/or processing speed [28]. These characteristics limit the ability to deploy traditional (i.e. conventional) security approaches with the IoT devices. For example, cryptographic mechanisms on these devices may need to be specially tailored to avoid the need for overly high processing requirements.

- **Scale:** The scale of an IoT system is extensive. An IoT system may handle billions (and potentially trillions) of devices, users and applications in real-time [29]. The number of devices, users, applications, services and their associations make it more complex when designing appropriate security measures for them.
- **Heterogeneity:** An IoT system is composed of numerous devices (e.g. sensors and actuators) and users. The devices may have different operating systems, hardware or technology compatibility (e.g. heterogeneous wireless communication technologies, protocols for networking, etc.). Furthermore, different hardware and software platforms facilitate the collection of data from heterogeneous IoT devices. As of now, there is no common standard for tagging and monitoring sensors in an IoT system [30].
- **Mobility:** The devices and users in IoT systems may be highly mobile in nature. While this is crucial for the overall performance of IoT applications in supporting realistic usage scenarios, it further complicates the provision of security solutions [31]. Some applications may be spread over multiple jurisdictions and/or change their jurisdictions within their life-cycle in the IoT system. They may lose network connectivity while roaming across the networks. In addition, resource availability and communication capabilities may vary throughout the networks.
- **Dynamic:** The interactions between entities may be many and highly varied. The number of entities that each device or user encounters can potentially be very large in number. Interactions may be very short in duration and pairs of entities may only interact once over the life-time of the IoT system. From a secure access control point of view, this is crucial when interactions may happen between entities that do not know each other's identities in advance [32].

2.2 Basics of Access Control

In this section, we aim at providing a basic introduction to access control technology. Our intention is to give an outline of the available access control mechanisms in general computing systems. This will help us to understand the concepts of access control mechanisms and its effective and valuable adaptation to IoT—the major focus of this book. Next, in Sect. 2.2.1, we present some definitions of access control followed by its working principle in Sect. 2.2.2. In Sect. 2.2.3, we provide a brief description of the available access control mechanisms, and discuss some available languages in Sect. 2.2.4.

2.2.1 Definition

Access control is a security mechanism that ensures the reliable access of resources by the authorized entities. Commonly, an access control mechanism describes how users and systems can communicate and interact with one another (or other systems

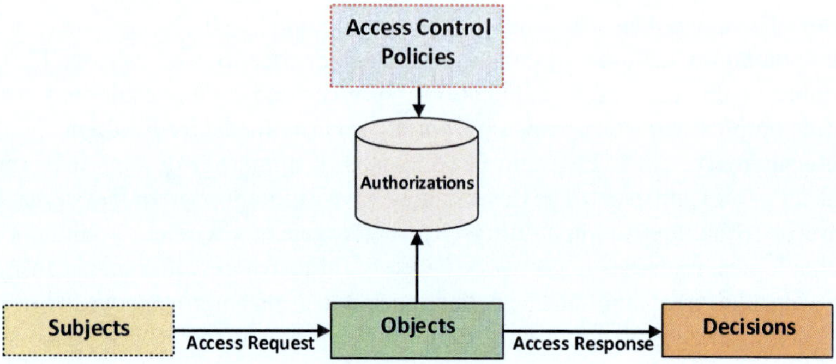

Fig. 2.2 Major components of an access control process

and resources) governed by the employed policies [33]. According to the Cambridge dictionary [34], access control is defined in two different perspectives. They are, (i) *"ways of controlling who can enter a building or area, usually involving electronic technology"* and (ii) *"ways of controlling who can see or enter information on a computer system"*. These definitions include the wider aspects of access control both physical and digital systems.

In [35], the definition of access control is that it *"constrains what a user can do directly, as well as what programs executing on behalf of the users are allowed to do"*. In [36] it is stated that *"access control determines what one party will allow another to do with respect to resources and objects mediated by the former, access control usually requires authentication as a prerequisite"*. In [37], access control is described as the *"security features that control how users and systems communicate and interact with other systems and resources"*.

Commonly, in a computer system, access control determines whether a subject (e.g. process, device, human user, etc.) is allowed to perform an operation (e.g. read, write, update, etc.) on an object (e.g. a database, file, service, etc.) based upon specific policies [38]. In other words, it governs who (e.g. a device or a user) or what (e.g. an application or a service) can view or use resources. In Fig. 2.2, we illustrate the main components of an access control process. Access control, in general, preserves the following properties [33]:

- *Confidentiality*: The information can be viewed by the authorized users and information must be kept private.
- *Integrity*: The authorized users can only write over information and information must be protected from being tampered with and altered by others.
- *Availability*: The information must be available upon request for use, which refers to the ability of a user to access a resource.

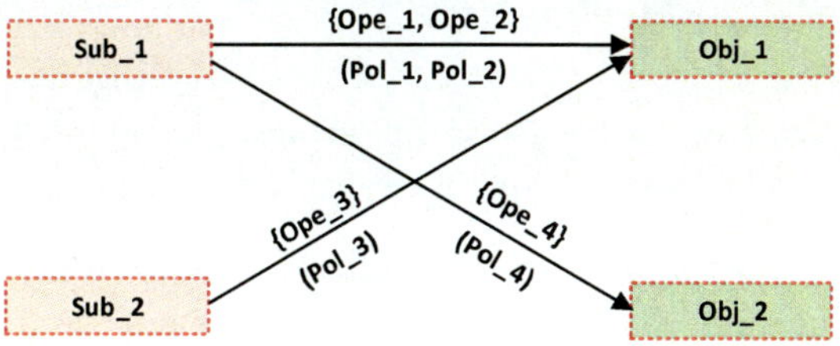

Fig. 2.3 A simple access control process

2.2.2 *Working Principle*

In Fig. 2.3, we illustrate a simple access control process. Where subjects (denoted as *Sub*) are trying to perform certain operations (denoted as *Ope*) over the objects (denoted as *Obj*) based on the specified access control policies (denoted as *Pol*). If the corresponding policies satisfy requirements, then the appropriate access will be granted otherwise the access will be denied.

Now let us consider a real-life example. Suppose a user Alice (i.e. subject in this case) wants to access online books (i.e. objects in this case) from University library. If the authorization credential satisfies the associated policies for that access, Alice will be permitted to access the desired resources. Eventually, there are various steps involved in an access control mechanism. When a subject requests an action to a certain object, the request must satisfy some conditions (i.e. attributes, contexts, etc.) based on the access control policies before the access permission is processed. In this case, we assume that the appropriate conditions are checked by the dedicated servers controlled by the University.

Potentially, access control can be achieved in two ways, *direct access control* and *indirect access control*. In direct access control, a subject can perform an action to an object directly. For instance, in the above example, Alice can access online books from the University library directly based on her student credentials. In an indirect access control, a subject can further transfer some of the access rights to another subject, who is then able to perform the intended action to the specified object. For instance, Alice wants to go for a holiday and wants to give some specific access permission (e.g. entering garden and garage) to her friend Bob to perform certain activities. To execute such access control, Alice can create and then transfer a 'capability' (also known as a *token*) [39] embedded with the specific access rights. This process of transferring access rights is called *delegation*. In general, access control can be done both physically and digitally. However, when we refer to the IoT, we refer digital access control. Throughout the book, when we refer to access control this will refer to a digital access control.

Depending upon the specific access control scenario and the information required for authorization to the resources, different access control architectures can be implemented [40–42]. However, this depends upon the system's requirements and the designer's choice.

2.2.3 Mechanisms

Different mechanisms are used for access control. In this section, we detail commonly available access control mechanisms for general computing systems.

- **Mandatory Access Control:** The Mandatory Access Control (MAC) model restricts access of any subject to an object based on the sensitivity level (e.g. secret, top secret, confidential, etc.) of access information of the object and authorization of the subject based on the authorization-rules [43]. These rules are governed by a central administration that is commonly known as security policy administrator. The sensitivity level is determined based on the system classification, configuration and authentication. Therefore, the security clearance of the subject and the security classification of the objects are bounded by the relationship of the subject-object pair and is stored in the security levels. In MAC, subjects cannot override or modify the security policies which allow the central administration to enforce strict security levels specific to their choices, where only the central administration can control all the tasks. As such, the end users have no control over any privileges. MAC is simple to enforce as it takes a hierarchical procedure (i.e. restrict the flow of information from more secure levels to less secure level) to employ access control policies to the resources. However, MAC is highly centralized in nature where for each access, a subject needs to obtain permission from the central administration. In this perspective, MAC relies on the system to control access and it does not provide a fine-grained access control to the resources in a large-scale system.
- **Discretionary Access Control:** The Discretionary Access Control (DAC) model allows access control based on the object ownership principle [44]. In other words, DAC allows access according to the discretion of the owner where the owner of the object specifies access control policies for each subject to access specific resources from specific objects. It allows an individual user complete control over any object they own. Unlike MAC, a subject that is allowed a discretionary access to a resource is able to transfer their access rights to another subject within the same group under the provision of an administrative policy. Interestingly, in DAC, access control is specified based on the identity of each subject or groups to which they belong. This is commonly achieved using Access Control Lists (ACL). The ACL is a tabular representation of an access control matrix of subjects that are mapped to specific resources belonging to specific objects. For example, the triple of $\{S, O, AR\}$ is an access control matrix where S is a set of subjects, O is an object and AR is a set of access rights. The set of access rights could be $\{read, write, execute\}$. Using DAC, a fine-grained access control can be provisioned but it is difficult with the growing scale of the system (i.e. where subjects and their access rights to specific objects increases).
- **Role-Based Access Control:** The elements of a Role-Based Access Control (RBAC) system are the users, *roles* and permissions—the central is the use of the *role* [45]. In RBAC, each user is assigned to a specific set of roles and permissions (i.e. authorized actions) are assigned to role(s) based on the policy decisions.

A role defines a function within the system in a hierarchical order. The relationship between users and roles is a many-to-many relationship mapping, where a user can be assigned to multiple roles and a role can have multiple users. Similarly, the relationship between roles to permissions follows a many-to-many relationship mapping. In RBAC, if the role of a user changes, then the corresponding permissions available to that user change. This means that user access can be adjusted by re-assigning role membership and without changing permission assignment. This provides greater flexibility in facilitating security administration for large organizations that need to manage their resources based on a user's responsibility and qualifications. Further, RBAC is well-suited for 'separation of duty' requirements, where all permissions are not assigned to one user for making a decision.

- **Attribute-Based Access Control:** In Attribute-Based Access Control (ABAC) access control permissions are assigned based on *attributes* [46]. These attributes can be seen as the properties that describe specific features of subjects, objects, environments, conditions, etc. For instance, name and age are subject attributes where time and location are environmental attributes. Policies are written to assign access permissions based on the attribute settings. Note that ABAC thus enables creation of access rules without creating explicit user to permission mappings. This can provide a significant degree of flexibility when compared to RBAC. In ABAC, access decisions can change between requests simply by altering the attribute values. There is no need to change predefined subject and object relationships.
- **Capability-Based Access Control:** In Capability-Based Access Control (CapBAC) access control permissions are assigned in the form of a *capability* (which can also be referred as a token, ticket or key) [39]. A capability can be defined as a communicable, unforgeable token of authority assigned to specific users for performing certain activities. A classic capability defines a resource and a set of access rights defining the operations allowed on that resource. In CapBAC, a user (i.e. subject) gets access to a resource (i.e. object) if the requested access (conditions) matches the contents of the supplied capability. The classical capability-based model was enhanced by Gong [47] who proposed the Identity-Based Capability System (ICAP), where a capability also includes the identity of the subject allowed the access. Identity-based CapBAC systems can avoid the centralization of standard RBAC and ABAC implementations.
- **Protocol-Based Access Control:** In a Protocol-Based Access Control (ProBAC) model, access control decision is examined in terms of the network and communication protocols used to deliver the authentication and authorization information. ProBAC depends upon the protocol choices of an access control architecture. For instance, access control based on Datagram Transport Layer Security (DTLS) protocol [48], access control framework based on Open Authorization (OAuth) [49], etc.
- **Hybrid Access Control:** In a Hybrid Access Control (HyBAC) model, two or more access control models are combined to serve system specific requirements.

In Table 2.1, we summarize the aforementioned access control mechanisms with a brief description for each of them.

Table 2.1 Various available access control mechanisms

Access control mechanisms	Brief description
MAC	Access control policies are determined by the system
DAC	Access control policies are decided by the object owners
RBAC	Users are assigned to particular roles and roles are mapped to certain permissions
ABAC	Access control decisions are determined based on certain attributes of the entities
CapBAC	Capability (also known as access token) is used for making an access control decision
ProBAC	Access control system is implemented based on protocols and frameworks
HyBAC	Access control system is made on two or more different access control mechanisms

2.2.4 Language

One of the challenging issues in access control is specifying and enforcing security policies that regulate the interactions between two parties e.g. subjects and objects. The basic concept of writing access control policies is the languages, commonly known as the access control policy languages [50, 51]. An access control policy language can be seen as a specific set of grammar, syntax rules (logical and mathematical) and operators that provides access control specifications combining subjects, objects and actions (i.e. the operations) [52]. This combines rules for authorization of certain actions. The operators use the attributes of the subjects and objects. With the increasing need for the flexible provision of the access control decisions, various access control policy languages are proposed. Next, we provide an outline of some commonly used access control policy languages.

- **eXtensible rights Markup Language (XrML):** It is used to define access conditions, rights and other related access control information related to a digital content [53]. In other words, it is a proposed language for standard Digital Rights Management (DRM). XrML is based on the standard XML (eXtensible Markup Language). In Fig. 2.4, we depict a simple policy expression. Where the *license* tag contains the *inventory* and *keyholder* tags, where the rights are managed and resources are given.
- **eXtensible Access Control Markup Language (XACML):** It is an open standard XML-based policy language for Web services. This is used to define and write general access control requirements followed by OASIS (The Organization for the Advancement of Structured Information Standards) standardization [54]. It has the ability to enhance the standard extension points for defining new functions, data types and policy combining logic.

```
<license xmlns:dsig='http://www.w3.org/2000/09/xmldsig#'>
   <inventory>
      <keyHolder licensePartId="123456">
        <info>
         <dsig:KeyValue> 123456 </dsig:KeyValue>
        </info>
     </keyHolder>
   </inventory>
</license>
```

Fig. 2.4 A simple XrML policy

We now provide an XACML-based policy specification to discuss how flexible authorization can be achieved. XACML implements ABAC as per NIST (The National Institute of Standards and Technology) guidelines and attributes are in the central role in XACML [55]. *Policy sets* are situated at the top of the hierarchy in an XACML document. Policy sets provide a means to combine other policy sets and (or) policies through policy combination algorithms. A *policy* is mainly composed of access rules and an indication on how to resolve the effects of these rules using rule combination algorithms. *Rules* are the primary elements that contain conditions regarding when an access shall be permitted or denied. This is where all the low-level logic regarding attributes are implemented and where the actual policy evaluation starts. An XACML rule contains four major parts:

1. **Effect**: Specifies the outcome of the rule when fully evaluated—either *Perimt* or *Deny*.
2. **Target**: Specifies the context in which this rule shall apply. For example, a rule specific to unlock doors requires at least an *action* attribute.
3. **Condition**: This is a constraint that shall be satisfied at the time of evaluation if the outcome is to take effect. For instance, it can constrain the rule to a set of specific resources.
4. **Obligation**: Describes a post-processing that shall be executed after a successful evaluation. For example, a capability shall be used before January 1st, 2019.

A simple policy document is illustrated in Fig. 2.5. According to the policy, when a request arrives at the PDP (Policy Decision Point), the target of each rule is evaluated sequentially and the first one that is applicable to the request is fully evaluated. Note, in Fig. 2.5, Rule 1 applies for read and write actions. Any other requests will be denied as per Rule 2. Rule 1 contains a condition which specifies this rule is imposed on a document `Doc`. The target and condition blocks must involve static attributes only. The policy described in Fig. 2.5 has one obligation which shall be fulfilled on permit (i.e. 01-01-2018 12:00:00).

- **JACPol:** JSON (JavaScript Object Notation) based access control policy language [56]. It is scalable and simple to use for policy specification in an expressive way. Note, JACPoL uses a fine-granular and hierarchically nested policy structure similar to XACML standard. It enforces a traditional ABAC system in a much

```
<Policy RuleCombiningAlgId="..:first-applicable" Version="3.0">
  <Rule Effect="Permit" RuleId="..:1">
    <Target>
      <AnyOf>
        <Match MatchId="..:string-equal">
          <AttributeDesignator AttributeId="..:act"/>
          <AttributeValue DataType="..#integer">read</AttributeValue>
        </Match>
        <Match MatchId="..:string-equal">
          <AttributeDesignator AttributeId="..:act"/>
          <AttributeValue DataType="..#integer">write</AttributeValue>
        </Match>
      </AnyOf>
    </Target>
    <Condition>
      <Apply FunctionId="..:string-equal">
        <AttributeDesignator AttributeId="..:trg"/>
        <AttributeValue DataType="..#integer">Doc</AttributeValue>
      </Apply>
    </Condition>
    <ObligationExpressions FulfillOn="Permit" ObligationId="time">
      <AttributeAssignmentExpression AttributeId="..:until">
          <AttributeValue DataType="..#string">
          [[TIME<01-01-2019 12:00:00]]
          </AttributeValue>
      </AttributeAssignmentExpression>
    </ObligationExpressions>
  </Rule>
  <Rule Effect="Deny" RuleId="..:2"/>
</Policy>
```

Fig. 2.5 A simple XACML policy

more flexible and fine-grained way. JSON is a light-weight standardized format for storing and transporting data as text over a network.

- **PTaCL:** It is known as the *'Policy Target and Composition Language'* [57]. It is an expressive policy-based access control language that provides authorization policy semantics which helps to understand the meaning of a policy for a certain request of attribute-based authorization policies. PTaCL is composed of the two sub-languages, named, Policy Target Language (PTL) and Policy Composition Language (PCL). PTL helps for target specification and PCL is used for policy specification. Note that the PTaCL is commonly used for the ABAC policy expressions that are used to evaluate access requests based on attributes associated with subjects, objects and actions.

2.2.5 *Cryptography*

Cryptography is a mathematical equation (or algorithm) that transforms simple data (i.e. ordinary plain text) into a complex and unreadable form (i.e. an unintelligible

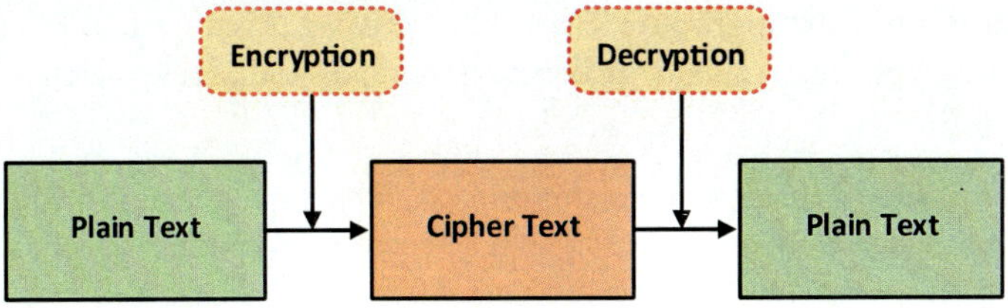

Fig. 2.6 An overview of cryptography process

text) where only the intended users can read and process it [58]. It can be seen as the process of storing and transferring confidential information from one entity to another in a way that the other entities are unable to view and modify the content of the information. In other words, the authorized users can only view the document. The core objectives of a cryptography process lies in various aspects in information security e.g. confidentiality, authentication, integrity, non-repudiation and to deliver anonymity to the communications.

The word 'cryptography' came from the Greek word *kryptos*, which means *hidden*. The origin of cryptography is considered to have been established as far back as 2000 BC. The first documented use of cryptography was seen by Julius Caesar (100 BC to 44 BC) who used to employ such a process using the Roman alphabet—one of the first modern ciphers. With the improvements of modern computing systems and the Internet technology, protecting data becomes a crucial issue and the use of cryptography plays an important part in this field.

Basic Terminologies: In Fig. 2.6, we illustrate an outline of a simple cryptography process. The most common process of cryptography is to encrypt the plain text and decrypt the cipher text using keys. Next we present the commonly used terminologies [59].

- *Plain Text:* It is the original message (e.g. a text file) that an entity wishes to share with another entity. It can be defined as an information that is pending input into cryptographic algorithms.
- *Cipher Text:* It is the encrypted message that cannot be seen or modified by anyone other than the receiver of the message. In other words, a cipher text is unreadable output of a plain text.
- *Encryption:* It is the method of transferring (i.e. encodes a message) a plain text into cipher text using encryption algorithms and keys. Note, this process takes place at the sender's side.
- *Decryption:* It is the method of transferring a cipher text into plain text using description algorithms and keys. Note, this process takes place at the receiver's side. Therefore, it is simply the inverse of an encryption method.
- *Key:* It is a numeric or alpha-numeric text (e.g. a string of bits) that is used for the encryption and decryption processes used by a cryptographic algorithm. Importantly, the length of the key is a major factor in a cryptography process.

Classification of Cryptography: There are two basic classifications of encryption algorithms. These are, (i) symmetric key cryptography and (ii) asymmetric key cryptography [60].

- *Symmetric Key Cryptography:* In this process, both the sender and the receiver use the same key, i.e. a single key is used for both encryption and decryption. The sender encrypts the message with a key and the same key is distributed to the receiver to decrypt the message. The keys can be shared using a secure communication channel between the two parties [61].
  There are various algorithms available to develop a symmetric key crypto system, for instance, Blowfish, DES, AES, etc. Blowfish is a well-known symmetric key block cipher, designed in 1993, that uses a 64 bit block size and a variable key length ranging from 32 bits to 448 bits [62]. The DES (known as the Data Encryption Standard) was popular in the 1970's adopted by the US government as an official Federal Information Processing Standard (FIPS). It uses a standard block size (i.e. the key length) of 64-bit. The DES was enhanced by the more advanced form of algorithm, the AES (known as the Advanced Encryption Standard). It is a symmetric block cipher that is capable of using 128 bit blocks with various key sizes ranging at 128, 192 and 256 bits, commonly known as AES-128, AES-192 and AES-256 [63].
- *Asymmetric Key Cryptography:* In this process, two separate keys (public and private keys) are used by the sender and receiver of the message. The receiver uses a 'public' key to encrypt the message and the received uses a 'private' key to decrypt the message. Note, for every public key there is a corresponding private key. The key used by the sender is known to everyone and therefore it is called the public key. On the contrary, the receiver's key is not shared to others and remains private. Asymmetric key cryptography is also known as the public key cryptography [64].
  There are various algorithms available to develop an asymmetric key crypto system, for instance, Diffie–Hellman, RSA, ECC (Elliptic Curve Cryptography), etc. Diffie–Hellman key exchange is one of the most important developments in asymmetric key cryptography. It is used to safely developing and exchanging keys between the sender and receiver of a message over an insecure (e.g. public) channel. This process allows exchange keys between two parties without the prior knowledge of each other to jointly establish a shared secret key over an insecure channel. In other words, this method helps two parties to share a secret key in a way that the key cannot be seen by observing the communication [65].
  RSA (known as the 'Rivest–Shamir–Adleman' algorithm—the developers of the algorithm) is one of the first and most popular crypto systems used in asymmetric key cryptography. The public and private keys in RSA is composed of two numbers, where one number is multiplication of two long prime numbers, along with an auxiliary value. The main idea behind RAS is that it is difficult to factorize a large integer. It can handle keys size typically of 1024 or 2048 bits long. This significantly improves the security in encryption strength in the cryptosystem [66].

ECC is another commonly used asymmetric key crypto system that uses algebraic structure of elliptic curves over finite fields. ECC uses elliptic curve equation to generate keys. Compared to RSA, it requires smaller keys while providing the same security, saying that a 256-bit ECC is equal to 3072-bit RSA key. The major motivation of making the shorter key size is the compatibility of the algorithm with low-powered and limited storage devices [67].

2.3 Access Control in the IoT

In this section, our aim is to examine the use of various access control mechanisms, discussed above (cf. Sect. 2.2.3), in the IoT. In an IoT system access control is employed with the context of the application and services that a user or device may wish to perform. The functionality of an IoT system is dependent upon the building blocks of the IoT and for each of them a different level of access control may apply [68].

2.3.1 Architecture

In [69], the authors present a discussion of the potential approaches to IoT access control architecture. They provide a categorization based on the externalized access control logic, collection of contextual information and location of their processing.

- **Centralized:** In this approach, the access control logic is externalized into a central entity (e.g. a Policy Decision Point located in a centralized server) that is responsible for making a decision based upon the authorization policies. Notably, this entity could be placed as a gateway that can facilitate communication to the devices or it could be seen as another entity located in a different location. The edge IoT devices act as passive entities and play a limited role as information providers. In this approach, the access control logic is located in a resourceful central entity. The users requesting for an access, need to be connected to the APIs provided by the central entity. However, this architectural approach suffers from the traditional drawbacks of a centralized architecture including a single point of failure and unavailability of contextual information.
- **Centralized and Contextual:** Contextual information (e.g. location, time, environmental status, etc.) related to the end-device are immensely important for a highly dynamic system like the IoT. In this mode of architecture, an access control decision is made based on the authorization policies located at a central entity as well as taking the contextual information from the edge IoT devices. Therefore, the edge IoT devices perform as an active entity. Note, in this case an extension of the end-devices is required by which the contextual information request can be transferred to the central system, e.g. the use of CoAP. Once again, this approach

suffers from the limitations of a centralized architecture. In addition, issues e.g. delays, overhead in communication between the devices and the central entity is crucial. Further, the end-to-end security cannot be achieved.

- **Distributed:** Notably, in this mode of architecture, there is no central entity involved. The access control logics are embedded inside the edge IoT devices. The devices are fully responsible to obtain, process and transfer information to other services or entities. This mode of architecture offers several benefits over the previously discussed architectures, e.g. the devices own full control to manage their information and the end-to-end security can be provisioned by the removal of the intermediate entities. While this approach outperforms more than the centralized approach, the obvious drawback is the resource constrained nature of the edge IoT devices to store and process the access control logics inside these devices.
- **Distributed Capability-Based:** In order to provide light-weight authorization mechanisms and to avoid the processing of complex access control policies, distributed capability-based access control architecture is proposed. It inherits the traditional characteristics of a CapBAC approach to a more fine-grained level. Inside the capability, access control permissions and other conditions of access (e.g. location, time, etc.) are embedded. Therefore, in this mode of architecture, when an entity receives the capability it already knows the level of access that has been granted to the service requester. This simplifies the authorization mechanisms involved by the centralized access control system and it addresses the resources-constrained nature of the edge IoT devices by avoiding the processing of complex access control policies. This approach also supports the delegation by distribution of access rights through capabilities.

2.3.2 Requirements

In an IoT system, it is essential to know the system specific requirements before employing an access control mechanism. There are several works that propose the requirements for IoT access control focused on architectures and the systems themselves. In this section, we list some of the common requirements when addressing IoT access control [70, 71]. However, we argue that these requirements may vary based on the characteristics and the specific needs of an IoT system.

- An access control architecture must consider the scale of the system. In other words, an access control architecture must consider an open environment where the number of devices, applications and services are not fixed.
- The access control mechanisms (and systems) should be easy to use and maintain for both the experts in engineering and the non-specialists.
- Access control should consider the nature of architecture when possible. The trade off between the distributed access control and centralized access control must be balanced. The client-based architectures [72] may be contemplated.

- The access control architecture should be flexible to adapt to the different contexts based on the system's deployment and consider the communities that share common attributes (e.g. location, mission, resource capability, etc.).
- The access control mechanism should consider the continual control where an access has a duration. It should be autonomous and self-contained.
- The access control system must support the dynamic attributes (both subjects and objects) and give the choice to update and/or change them at anytime, and users can personalize event-based scenarios. Systems based on concrete identity of a *thing* can be avoided. Towards this, identities of access *things* can be made attribute-centric.
- The access control system must reinforce the integration with various third parties. Towards this, the architectural design should be taken into account, an open framework to accommodate compatible components.
- Access control in IoT must be light-weight in essence, given the resource constrained nature of the IoT devices. The traditional cryptographic methods introduce a heavy computation overhead for the device involving complex key distribution and flexible data management.

2.3.3 Existing Works

In this section, we examine three commonly used access control mechanisms, namely, RBAC, ABAC and CapBAC (outlined in Sect. 2.2.3) for the IoT that are significant to this research.

- **Access Control in IoT Based on RBAC:** There are several approaches that use RABC to address access control issues in IoT. For instance, Kulkarnin and Tripathi [73] discuss a *Context-Aware RBAC (CARBAC)* model for pervasive computing systems. The motivation of the model is to build RABC models and systems for pervasive computing systems where context information is provisioned for making access control decisions. In this model, users' memberships are mapped to the roles and permissions are executed by the role members with the context-based dynamic integration of services, in a specific environment. The contexts can be a user's physical location, the device being used, network in which the devices are connected or the user's current activities. The use of context facilitates the revocation of access rights by failing specific condition within a given context. The proposed CARBAC is dynamic in nature when making an access control decision within a context.
  Zhang and Tian [74] discuss an access control approach that uses context information and RBAC for large-scale systems like the IoT. This is an extension of the traditional RBAC model where operations on objects are converted to services and the permission for accessing resources by the subjects are given based on a set of contextual information that are collected from the system. Services are referred to as a set of functionalities that a device should offer to

the system. In this model, a centralized security administrator assigns appropriate permissions to roles on the basis of the characteristics and context of physical objects, and specify an appropriate range of the users according to the function of the role. In general, the context is described as any information that can be used for characterization the situation of an entity. The model is composed of the following components: $\{U, R, P, C, Ser\}$ they are: users, roles, permissions, contexts and services respectably. An access decision can be checked on the following set, for example, where a user (e.g. a student Alice) can access a University parking spot between 9am and 5pm on weekdays as follows $\{Alice, Student, ParkingInWeekdaysOncampus, 9am{-}5pm, ParkingCar\}$. With similar objectives to [73], the proposed approach in [74] discusses the integration of RBAC in context information for dynamic systems. However, the model discussed in [73] is more flexible than [74] in various ways, e.g. some context-based conditions must be satisfied before admitting a user to role, and also for granting a user-role membership and personalized role permission is introduced which allows different role members to access different active space services within a given context.

Zhang and Parashar [75] discuss a *dynamic RBAC* approach for pervasive applications e.g. for a smart home. Similar to the concepts of [73, 74], the proposed model is able to make access control decisions dynamically based on the contextual information. It extends the traditional RABC model and dynamically adjusts the role assignments and permission assignments to the specific roles based on context information. In this model, each user is assigned a role subset and resources have permission subsets for each role that will access the resource. However, unlike [73, 74], in this case, a state machine is maintained for delegated access control agents at the subject (to navigate the role subset) and the object (to navigate the permission subset). The state machine maintains the role-permission subset to react to changes within the given context.

Kalam et al. [76] discuss an access control model from an organizational point of view namely *Organization Based Access Control (OrBAC)*. This is an extension of RBAC, where the role is defined by a group of users performing a particular task within an organization. In a common access control mechanism, e.g. DAC, the basic relation is constituted as a triple $\{subject, action, object\}$, in OrBAC, this is abstracted to a more generalized level, which consists of the triple $\{role, activity, view\}$. Where, role is a set of subjects, activity is a set of actions and view is a set of objects, within a particular context. A subject can be either an active entity (i.e. a user) or an organization. In this model, the specification of security policies are parameterized by the organization which provides flexibility in handling several access security policies that are associated with various organizations. Unlike, [73, 74], this model is centralized in nature when considering the role-membership assignment for a specific organization. However, this model is not restricted to permission and includes the possibility to specify prohibitions, obligations and recommendations that apply to subjects, objects and actions.

Pasquier et al. [77] enhance the concept of OrBAC [76] to the IoT and propose a *Smart Organization Based Access Control (Smart OrBAC)*. This attempts to extend

OrBAC for an IoT environment with a set of security and performance requirements that respects the characteristics and the constraints on the smart objects in an IoT system. Unlike the previous approaches, e.g. [73, 76], this approach addresses the authorization and access control issues in the context of distributed, cross-domain systems that potentially consists of resource-constrained IoT devices that perform autonomously without any direct human interventions. Compared to [73], where interactions between various devices are considered between various organizations, this model focuses on a single constrained device and its communication between several other devices from different organizations or domains. Similar to [74], this model facilitates a distributed-centralized approach where authorization decisions are made upon the local conditions within a given context. Moreover, Smart OrBAC model is conceived through an abstraction layer design. It partitions access control process in different functional layers of an IoT architecture, given the fact that every device is not constraint uniformly to every layer. Unlike, OrBAC [76], this model addresses the concept of collaborative interaction within a given context. This is handled at the 'collaboration layer' where two organizations can communicate seamlessly. The collaboration is done by making a prior agreement between the involved organizations where both of them can jointly define access control policies.

Freudenthal et al. [78] present a *distributed RBAC (dRBAC)* model. This is an extended version of RBAC that supports collaboration among large-scale coalition environments. The traditional RBAC systems depend upon a central trusted computing base administered by a single authority. dRBAC extends traditional RBAC to support for multiple trust relationships for access control using systems like Simple Public Key Infrastructure (SPKI) [79]. This leverages the features of RBAC and trust-management systems to create a system that offers both administrative ease and a decentralized scalable implementation. It is a decentralized access control mechanism for large-scale systems that span over multiple administrative domains. dRBAC utilizes Public Key Infrastructure (PKI) based infrastructure for managing identities of entities to establish trust over the multiple administrative domains. Roles are defined by controlled activities. Unlike [77], in this model, permissions are assigned and distributed across domains in terms of delegation.

Liu et al. [80] discuss an access control model for an IoT system based on ECC and RBAC. RBAC is enhanced for authorization for IoT devices using their (i.e. device's) particular role and applications that are associated in IoT systems. ECC is used for key establishment during entity authentication and RBAC is used for specifying access control policies. Unlike, [74, 77], this paper provides a detailed security analysis of the system. However, this paper only outlines the approach and how this approach will work in real-life IoT scenarios is not discussed. Further, no implementation is given. In [81], Liu et al. present an access control architecture for resource sharing in large-scale IoT systems employing RBAC. The subjects are given specific roles and roles are assigned to permissions according to security policy specifications of a central administration authority. Like [77], this model also focuses on collaborative environments for access control between subjects

and objects. Unlike, [77], this model sets up a formal model for RBAC safety policies.
Other approaches e.g. Jindou et al. [82] and Barka et al. [83] use RBAC for the Web of Things (WoT) [84]. In the WoT, physical *things* can be accessed and controlled via the Web. For example, people can share a collection of *things* e.g. temperature measurement sensors, air conditioning machines, music players, monitors and lamps in a smart home, notification of an event, etc. These approaches enhance RBAC to control access to *things* on the Web. Jindou et al. [82] integrates SNS (Social Network Structure) into RBAC to allow dynamic policy of access control to support flexible access control on IoT *things*. The SNS enables users to share *things* with other users who they know and trust based on user profile and social links. Similar to [82], proposal [83] presents an architecture that utilizes the features of the RBAC for maintaining access control policies for the WoT, and cryptographic operations are used to enforce such access control policies. The proposed architecture integrates properties of RBAC (e.g. data abstractions) to specify the access control policies to the WoT. Unlike [82], this proposal does not include dynamic contexts of the environment and the SNS structure when making an access control decision.

- **Access Control in IoT Based on ABAC:** There are several proposals that discuss the use of ABAC in IoT. These models take into consideration different attributes, for example, user's name, location, context information, proximity, behaviour or even activity. For instance, Zhang and Liu [85] present an ABAC model that provides for fine-grained access control for IoT systems. The proposed model allows permissions to be assigned to a user for accessing resources based on user attributes, resource attributes, environment attributes and current tasks. It introduces the use of service-oriented computing to address IoT access control using contextual information and directly interacting with the objects. This model allows for the dynamic characteristics of the *things* in an IoT system, supporting the principle of least privilege and dynamic separation of duties. However, it requires policies to be written on an individual user basis.
Similar approaches for using ABAC for the IoT can be seen in [85] and Bezawada et al. [86]. They apply ABAC for securing IoT-enabled smart home environments where diverse and independent computing devices provide many services to the users. Various entities in the home IoT environment are categorized as either subject or object, and then assigned corresponding attributes. The attributes are described as to the formal entities in ABAC e.g. subject, object, resource, and user. These attributes are obtained by real-life lab testing of device characteristics and from manufacturer specifications. Then appropriate policies are written for different categories of subjects and objects within the home IoT environment. This is a conceptual model and no implementation is provided. Unlike [85], this approach considers a closed IoT system (i.e. a home environment).
Unlike [85, 86], that address access control in core network level in IoT, Ye et al. [87] present an access control model for the perception layer of an IoT architecture. In this model, ABAC is used for access control decisions enhancing the fine-grained access control aspects of ABAC implementation for complex system

or dynamic extension of large-scale users. For mutual authentication and secure key establishment between the users and IoT devices, ECC is employed. Mutual authentication ensures the security in the communication between the users and edge IoT devices.

Touati and Challal [88] present an *Activity-Based Access Control* model for IoT. Here the term activity is referred to as the context-information (as an attribute) that is taken into consideration for an access control decision. This model allows a fine-grained and context-aware access control that takes into consideration user's context evolution and leverages the advantages of an ABAC system in an IoT system. The model is implemented in a finite-state machine and an asymmetric encryption mechanism namely Ciphertext-Policy Attribute-Based Encryption (CP-ABE) is used to achieve a real-time access policy adaptation following users and system context evolution.

Sciancalepore et al. [89] present an ABAC scheme for federated IoT systems. The proposed scheme features distributed and decoupled mechanisms for authentication and authorization services in heterogeneous and federated IoT systems. With ABAC, this scheme also leverages the advantages of token-based authorization techniques e.g. using JSON Web Tokens (JWTs). The access control decision is based on policy specifications by the standard ABAC engine and the decision is transformed to the users via web tokens. This scheme integrates several IoT platforms that belong to heterogeneous resources. All resources are registered with a trusted mediator which offers mechanisms for enabling platform interoperability and distributed resource access. No implementation is given to support the scheme in real-world IoT scenarios.

Similarly to the concept of using context information as discussed in [88], Lang and Schreiner [90] present an *Proximity-Based Access Control (PBAC)* model for IoT. However, unlike [88], in this case, proximity is used as an attribute. The access control decisions are made based on the proximity between attributes associated with two (or more) entities. The proximity (e.g. the distance between the subjects and objects) is used as an attribute with an extended ABAC. These enhanced access control approaches can implement flexible, proximity-based, dynamic, contextual access for large-scale dynamic systems e.g. the IoT. This model derives from ABAC and Model Driven Security (MDS) [91] to express and enforce security and privacy requirements of the access control. The proximity is measured by a distance calculation function. Notably, the proximity is not only limited to distance, it can hold other attributes as e.g. geospatial location, organizational, operational, temporal, business process, etc. The model provides more controlled ability to define and implement better user privacy but it also suffers from a lack of available attribute information sources which is a challenge in this access control model.

The UCON (Usage CONtrol) model [92] encompasses traditional access control, trust management and digital rights management (DRM). It allows authorization involving both rights-related and obligation-related authorization rules as well as conditions. UCON is a generalization of access control where the authorization process is based on the subject attributes and object attributes. These attributes can be identities, security labels, properties, capabilities, etc. Along with these attributes authorizations, obligations, and conditions need to be evaluated for making a usage decision [93]. In a traditional ABAC model, the subject and object attributes can only be changed before the access request. However, in the UCON model, these attributes could be changed not only before the access request but also during and after the access request and this will affect the access granted. This change will affect the permission decision in the subject's next access behaviour. Guoping and Wentao [94] use the UCON model for access control in IoT to meet the needs for security authorization and control. The authors propose a framework based on a network layer design. A mapping of UCON abstractions and IoT entities is provided in detail. For example, an entity (e.g. subject(S)) is defined as the Device(D). The attribute(S) of UCON in IoT for the subject represented by the Att(Device). Similarly, the condition(C) of UCON in IoT is decided by the access control policies e.g. trust value. The oBligation(B) of UCON in IoT is according to the needs of the wireless sensor network. The Authorization(A) of UCON in IoT is set by the needs of usage control, and decided by the device and the service. An outline of theoretical experiments is given with an assessment model based fuzzy theory, however, no proof of the concept prototype is presented. Therefore, the proposal [94] discuss an access control model based on fuzzy theory the abstraction of UCON model. However, this approach provides a limited discussion on the effectiveness of using UCON for IoT, and no implementation is supported.

- **Access Control in IoT Based on CapBAC:** There are several access control mechanisms that are derived from the fundamental concept of CapBAC and employed in IoT systems. For instance, Gusmeroli et al. in [39, 95] discuss a CapBAC approach for the IoT. CapBAC systems use capabilities to manage access control processes. Recall, a capability can be defined as a communicable, unforgeable token of authority. This refers to a value which can be associated with an object (uniquely identified) with some set of access rights to obtain the resources and required permissions. The access control permissions are embedded inside the capabilities and the access decisions are made inside the edge IoT devices at the time of an access. These approaches facilitate systems that enterprises, or even individuals, can use to manage their own access control processes to services and information. These systems are implemented so that capability checking is decentralized as the capabilities are signed by the capability issuers. These signatures can be checked by the edge devices. This is a major advantage in an IoT system. Another advantage of the use of a CapBAC approach is the employment of an encrypted capability chain for secure capability delegation which is immensely important in highly dynamic systems like the IoT. While the proposed delegation models overcome the shortcomings of the centralized system, it does not solve the problem of delegation at a fine-grained level of policy management. The system-generated capability can be delegated to anyone by the owner of the capability. The only restriction is the depth of delegation.

Similar to [95], Hernandez-Ramos et al. [69] discuss a *Distributed CapBAC* model for IoT. In this model, the traditional CapBAC is enhanced with the distributed approach in which the edge IoT *things* are capable of making authorization decisions by themselves (at the time of an access) without the need of a centralized server. The authorization decisions are made based on the available local conditions within a given context offering context-aware access control which is significant for an IoT system. The access control decision is done at the presentation of a capability to the edge IoT *things*. This approach presents a cryptographic solution that supports the standard certificates based on ECC against insider threats through the Distributed CapBAC. A highly optimized version of Elliptic Curve Digital Signature Algorithm (ECDSA) is implemented inside the *thing* (i.e. the resource constrained devices) for ensuring an end-to-end authentication, integrity and non-repudiation which does not involve any intermediate entity for computing computation. Proof of the concept prototype is detailed and experiment results are depicted to support the feasibility of the model in an IoT environment. This approach is superior to a completely centralized system that suffers from a single point failure, and addresses the issues of scalability and end-to-end security, which cannot be fulfilled by traditional access control approaches discussed above.

The work in [96], is similar to the research presented in [69]. The concept in [96] also uses the distributed nature of an IoT infrastructure. The proposed model is based on the design of a light-weight token that is used for accessing CoAP Resources. An optimized implementation of ECDSA is employed inside the edge IoT devices where the devices themselves are capable of performing the access control decisions. In both of these approaches, the authorization tokens are written in JSON and are sent to the target edge IoT resources (i.e. the *things*) using CoAP. Moreover, these approaches are developed based on the public key cryptography which is not an ideal base for constraint IoT devices.

Similar to [69, 96], where access control decisions are made inside the edge IoT devices, Hernandez-Ramos et al. [97] present a set of light-weight authentication and authorization mechanisms in order to support smart objects during their life cycle. Unlike [69, 96], in [97], the authorization and authentication mechanisms are integrated and extended with other standard technologies that address the various security planes in an IoT device life cycle within the scope of an ARM (Architectural Reference Model) compliant security framework [98]. The ARM is built to optimize the interoperability issues between isolated IoT applications to create a global ecosystem of services under common understanding that is achieved through a set of specific tools and guidelines.

In [99], Zhou et al. discuss a flexible and fine-grained access control approach for the IoT. The motivation of this study is to provide granularity and techniques for supporting large scale operations to make flexible interactions between the entities in an IoT system. A decentralized strategy and data centric network techniques are used to enhance the model. This proposal only outlines the principles of the proposed approach and gives an indication of using CapBAC for flexible access control and their management in a decentralized way. In [100], Zhou et al. extend the approach of [99] with the complete system design and its implementation. The approach consists of a three-tier architecture (namely, back-end, resource-rich objects and resource-constrained objects) to provide centralized policy management and distributed execution of access rights for large scale enterprise environments. When users need an access to a resource, the users must register themselves to the back-end server in order to get a capability for the specific resources. This proposal also highlights the essence of access right delegation for the resource constrained IoT devices. A detailed system design and implementation is provided to support the proof of the concept architecture. Similar to [97], capabilities are used for access to resources and once the capability is obtained, the access is no longer involved with the central entity (i.e the back-end server in this case). However, unlike [97], in this approach, it is assumed that the objects are largely static once installed in the environment. Also, it is assumed that the back-end server, subject devices and objects are synchronized in a timely manner. The subjects and the objects are characterized by their unique identity (for instance, with a device id or a serial number) or with the predefined attributes that are stored in the central entity during the registration of the subjects and the objects.

Anggorojati et al. [101] discuss a Capability-based Context Aware Access Control (CCAAC) model for a federated IoT network. The proposed access control model is based on a centralized administration that is maintained by heavy-weight Web-based applications. This model also considers the context of the environment when processing a delegation request. While the model can provide a fine-grained policy control of delegation, it is highly centralized in nature, which is not an ideal choice for the IoT. The proposed model incorporates identity-based capability and dynamic context information.

Xu et al. [102] discuss a *Federated CapBAC* approach for access control in IoT. In this model, traditional CapBAC model is used for delegating access rights over a federated IoT networks spread across multiple jurisdictions. This is achieved by means of the capability propagation and incorporating the context information within the delegation. Once again, this model is highly centralized. It brings an identity-based capability management strategy which involves registering, propagation and revocation of the access authorization. Unlike [69, 96], this model addresses the issues of capability revocation and delegation considering the granularity and context-awareness of an IoT system. Similar to [39], this model provides a light-weight solution for access control for IoT edge devices, however, unlike [39], this model considers a multiple federated domains for capability propagation.

Mahalle et al. [103] discuss an *Identity Establishment and CapBAC (IECAC)* scheme for the IoT. The protocol for capability verification by optimized and scalable ECC authentication and access control protocol. The protocol is divided into two parts: a one way authentication, and mutual authentication and integration with traditional CapBAC solutions. This model improves the security assertions of Gong's model [47] for creation and propagation of capability.

In summary, the massive scale of the integration of heterogeneous devices and services in an IoT system means that none of these aforementioned access control approaches (e.g. RBAC, ABAC and CapBAC), in isolation, can achieve efficient management of access control policies. Recall, RBAC provides rights to specific roles and users are made members of appropriate roles, rather than giving rights directly to the users. This simplifies policy management. However, to explicitly identify and assign users to roles is difficult in a dynamic and large-scale system e.g. the IoT. RBAC systems are complex and not easily implemented in constrained devices. RBAC is also inherently static (with role membership needing to be defined a priori), and typically highly centralized in their implementation and the adjudication of access. This makes an unmodified implementation of RBAC unsuited for IoT systems. ABAC makes access control decisions based on the 'attributes' of system entities. ABAC can also provide a context-aware approach based on properties e.g. location and time. ABAC can support fine-grained access control, but at the cost of significant complexity, both in number of policies and details of policy expressions. CapBAC approaches provide fine-grained access control by supplying a capability which precisely specifies the access allowed to a resource (e.g. a *thing*). While capabilities can provide a fine-grained approach to access control, defining the policies that will provide a capability to each user for every resource they may access is not scalable without the addition of significant machinery for policy management.

Access control for IoT systems needs to be scalable, flexible, usable and recognize the inherently decentralized nature of such systems. Moreover, the intrinsic features of traditional access control approaches may be difficult to implement within the resource-constrained IoT devices. There is a requirement that, whatever access control mechanism is employed, it should be usable as well as sufficient to protect the privacy, integrity and confidentiality of the system and its components. In general, in an IoT system, information can be available and open to everyone. However sensitive and confidential information should be protected allowing only authorized users to control and manipulate that data. The issues noted above around the existing approaches to access control mean that further consideration of access control for the IoT is merited.

2.4 Representing Identity

The concept of identity has been widely discussed in industries and academia. With the emergence of the Internet and various communication technologies this became an important issue. In computing systems, an identity is the basis of the assignment of privileges that are verified through dynamic data. To represent an identity, information and communication systems generally use various identifiers for identification and data transmission over a communication channel [104].

Identity represents the fact *who it is*. Identity can be an entity's physical parameters (e.g. name, age) or digital information (e.g. credit card number) that uniquely identify an entity within a given context. The entity can be a human, device or an organization. There are various representations of identity, for example, using attributes, claims and partial identities. It is discussed in [105] that *"authentication is the binding of an entity to a subject"*. Authentication governs the processes of obtaining 'authentication information' from an entity, analyzing the data and ensuring whether the information is associated with that entity. In digital systems, this highlights the fact that a computer system must store certain information about that entity. This information can be represented as the identity of the said entity within a given context. In other words, an entity must bind to a specific identity that uniquely represent that entity within the given context.

As noted earlier the IoT will include a very high numbers of users, *things* and resources. There will be a need to identify these elements. In our particular case, as the subjects and objects of policy. Various characteristics of the IoT mean that a simplistic approach to identity will be insufficient. Addressing each entity individually will require a vast number of policies and reduces finesse in policy expression while increasing the number of polices that need to be created and managed. A more sophisticated approach to identity will allow multiple entities to be referenced by a single identity. This situation is further complicated by the nature of the IoT, e.g. device diversity, broad scope of interactions, different communication mediums and dynamic characteristics of the devices, services and applications. In particular, it is difficult to ensure in advance what and when an entity will interact with another entity. There is a need for addressing IoT identity in a systematic way that will help to enable identity management for billions of *things* which can access and be accessed in a heterogeneous environment. This will in turn assist to develop a scalable, dynamic and flexible IoT access control architecture. Identity, its meaning and expression needs to be addressed in arriving at a flexible solution to access control in the IoT. In Chap. 5, we discuss the notion of identity in the context of the IoT in a comprehensive and systematic way.

2.4.1 Identity Establishment

The identity establishment process determines the identity of an entity based on some identification factors. This enhances the fact that the entity must provide certain information that helps to enable the system to confirm the identity of that entity. Identities are created and managed by a management authority, known as the identity manager or identity provider. Note, only providing an identifier as an identity information is often insufficient to prove a claimant. Identity establishment is closely associated with the entity authentication and authorization.

- *Entity Authentication:* Recall, authentication provides an assertion to verify an entity. An authentication procedure comes in terms of one or more of the following [105]:

  (a) *What the entity knows:* In this case, an entity possess knowledge which identifies the entity. Commonly used types are password or any secret information that belongs to the entity.
  (b) *What the entity has:* In this case, the systems verify certain information that an entity holds. Commonly used forms are a security token, a key fob or an identity card where the information is securely stored.
  (c) *What the entity is:* In this case, an entity's identity is verified using some unique features. For instance, biometric information (e.g. fingerprints, retinal and facial characteristics) that is unique to an entity.

(d) *Where the entity is:* In this case, an entity's current geographical location and time in a specific terminal is considered for identity verification process. Therefore, this process is context dependent.

- *Entity Authorization:* Although authentication provides reliability between the communicating parties, authorization is needed to provide secure access to private and confidential information only to specific users. Recall, authorization is the procedure of specifying access rights to certain resources based on the employed policies. To this end, identity is crucial in an access control mechanism. Note, in Sect. 2.2.3, we provide a detailed discussion of various access control mechanisms.

To provide increased security to a system, the activities of an entity (e.g. interaction time, resource accessed, etc.) are tracked and recorded. This process is known as *accounting*. Authentication, authorization and accounting are often provided by a dedicated server, called the *AAA server*.

2.4.2 Identity Management Models

Identity management is the process of managing, controlling and maintaining of identities [106]. The process consists of one or a set of events for the creation, usage and the formulation of identities for the said management in a particular domain. The process of identity management can be done both physically and digitally.

The core components in an identity management system are the clients (i.e. the users), service providers and identity providers [107]. The service providers allow services to the clients. They also store the clients identifies and authorization credentials for authentication purposes. Finally, the identity providers are responsible for registration, verification of a client's credentials and the creation of identities. The commonly used identity management models are [108]:

- *Isolated Identity Management Model:* In this model, an entity must register to a service provider. Each entity possesses an identifier for accessing an isolated service. This model is mostly flexible for the service providers where the service providers act as both the identity and credential providers for their clients. In other words, the service providers act both as the service provider and identity provider, where identity storage and entity's operations are performed by a signal service provider.
- *Centralized Identity Management Model:* This model is used for most identity management systems. This model separates the functions of service provider and identity provider. The identity storage (of an entity) and its authentication are performed within the identity provider server. In this model, a single identifier and credentials are used by each service provider. In other words, all service providers use the global unique identity provider. For example, a PKI based infrastructure where a Certificate Authority (CA) issues certificates to the users.

- *Federated Identity Management Model:* In this model, agreements for authentication are established between the service providers (that trust each other) so that the identities from different service providers in specific identity domains are recognized across all domains. An entity belongs to a particular service provider is able to access services provided by a different service provider over a *federated-group* using a single identifier. In other words, when an entity is authenticated to one service provider using one of its identifiers, subsequently the entity is considered as an identified and authenticated entity by the other service providers. This simplifies the account management problem between two different service providers. However, from an entity's perspective, it may still need to remember multiple identities and credentials.

2.5 Approaches to Delegation

Delegation is an important part of access control mechanisms. Recall, access control governs who or what can view or use resources by the allocation of rights based on the policies specified. Initially rights are bestowed by the owner of a resource. Delegation allows entities to transfer those rights to other users. The scale of and dynamic nature of the IoT means that it requires flexibility in its approach to policy management. Such systems may generate massive amounts of data, much of which may be sensitive (e.g. government or health related information). As we indicated earlier, protecting this data from unauthorized users and services requires proper access control. Enforcing appropriate access control is an important part of securing these systems. An important feature of access control systems to deal with flexibility is delegation and it has been included in existing proposals for IoT access control. However, delegation is usually framed by explicitly identifying the recipient of the delegation. As noted under the discussion of identity, such an approach is not necessarily suited to general IoT usage. The nature and implementation of delegation within IoT access control deserves further examination. In Chap. 6, we provide a detailed discussion on delegation of access rights in the IoT.

2.5.1 Definition

According to [109], delegation *"is the assignment of any authority to another person (normally from a manager to a subordinate) to carry out specific activities"*. Fundamentally, a delegation states that how an active entity (e.g. user, device, etc.) can transfer and grant some of their permissions (i.e. access rights) to one or more entities for accessing one or more resources. Therefore, delegation can be seen as a process by which an entity can transfer one or a set of access rights to one or a set of other entities. In a delegation, the entity that transfers the access rights is called the *delegator* and the receiving entity is known as the *delegatee*.

2.5.2 Architecture

Next we use an example scenario to demonstrate the delegation process. Let us assume Doctor A and Doctor B work for the same health care provider, i.e. are located in the same network domain (e.g. working for the same health care provider). Alice is under the care of Doctor A. Doctor A is going to be absent for a short period of time and Doctor B will provide care in Doctor A's absence. Changing the underlying policies, and changing them back on Doctor A's return, is inefficient. Delegation provides a simpler and flexible short-term option. The following are some commonly used delegation architecture.

- **Centralized Checking on Issuance:** In a centralized issuance delegation architecture (e.g. [110]), delegation is handled by a central administration entity. In this case, the *delegator* contacts the central administration with the request for a delegation to the *delegatee*. This request will include the rights to be delegated and possibly other information, e.g. the requested duration for the delegation. The central administration will typically hold policies which will govern the delegation and enable a decision as to whether the delegation is allowed. If it is allowed, the *delegatee* is granted the delegated rights.
  Figure 2.7 illustrates the centralized issuance delegation architecture. In this case, Doctor A requests the central system administration to create a delegation for Doctor B. If the relevant policies are satisfied, then the central system grants the delegation to Doctor B. Now, Doctor B is able to see relevant medical details of patient Alice and provide care. While the centralized approach allows a fine-grained, policy-based, approach to the management of delegation, it requires the active participation of the central administration entity. This will not always be practical in the IoT context.
- **Centralized Checking on Access:** In this case, a *delegator* issues delegated credentials directly to a *delegatee* (e.g. [111]). In contrast to the previous approach, the delegation is only checked when the *delegatee* attempts to exercise their delegated rights. As shown in Fig. 2.8, when the *delegatee* wishes to use the credentials they are checked by a central component of the system (e.g. an authorization authority)

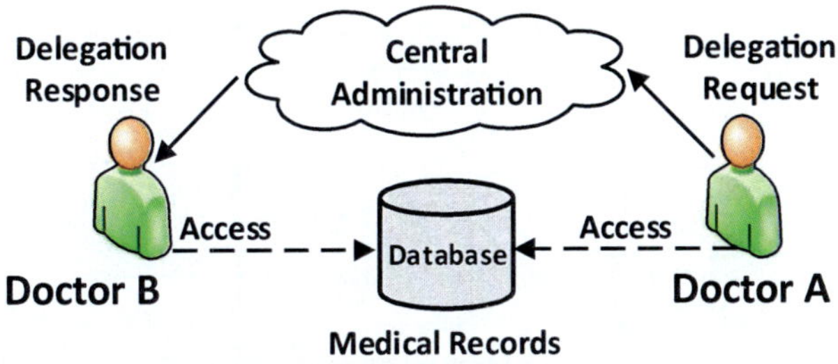

Fig. 2.7 Centralized checking on issuance of delegated access rights

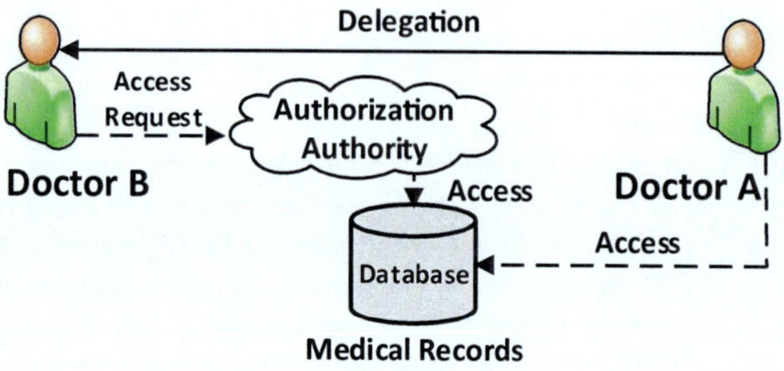

Fig. 2.8 Centralized checking on use of delegated rights

before access to the resource is allowed. Like the previous approach, the centralized component may contain fine-grained policies that govern the delegation.
Centralized systems have the advantage of easily allowing a policy-based approach to delegation. However, a major disadvantage is that they require the active participation of a central component either when the delegation is authorized or delegated rights are used. While this is reasonable in a wide range of application scenarios, it is not suitable for highly dynamic systems like the IoT.

- **Distributed:** In a distributed delegation architecture (e.g. [39]), credentials for access (e.g. capabilities) are issued to the users. These credentials are checked on access requests. Like the previous approach a *delegator* issues delegated credentials directly to a *delegatee*. As shown in Fig. 2.9, the *delegatee* presents the credentials on access. The service provider (e.g. *things*) checks that the delegation is genuine as the delegated credential either includes or is accompanied by a complete authorization chain to enable validation to occur.
Distributed approaches to delegation have the advantage of not requiring the involvement of a central component. This assists in distributing the management of the authorization process, as advocated by [39]. However, they do place extra requirements on the service providers. This is a well-known trade-off between centralized and distributed approaches. On the one hand, too much decentralization risks a loss of control and vulnerabilities occurring in the independent components. On the other hand, too much centralization risks the creation of unscalable

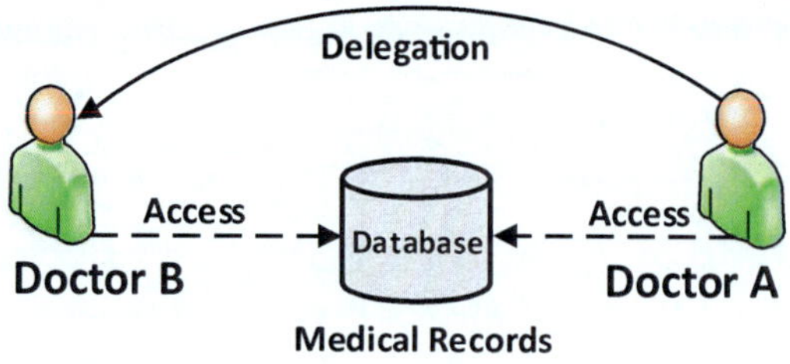

Fig. 2.9 Distributed delegation of rights

solutions. Distributed approaches also have difficulty in providing fine-grained, policy-based control over delegation, typically leaving all (or at least most) decisions over whether or not a delegation is valid to the *delegator*.

Other delegation approaches are possible, e.g. policy-based delegation, rule-based delegation, etc. However, for the purposes of the IoT, they can be placed in the above mentioned architectures.

2.6 The Notion of Trust

Trust is one of the most important features in our everyday life. Trust provides many practical benefits including safe and inexpensive basis that enables cooperation between two or more entities when uncertainty and risks exist [112]. Commonly, a question can arise, whether we can trust an entity and if so in what aspect. This is fundamental in every aspect of our lives where we need to adjust decisions accordingly. Let us consider the following situation where Alice trusts Bob for service X, but not for the service Y. So Bob is trustworthy in the context of service X but not in the context of service Y. Trust is context sensitive, subjective and may vary in different ways based on the social issues.

In this section, we first discuss the concept of trust from a general point of view and then outline the trust modeling process. Next, we present two mechanisms for trust evaluation that are related to this research. Finally, we provide an outline to two categories to calculate trust.

2.6.1 Trust Concept

The notion of computational trust dates back to the early nineties. In the past decades, there have been many areas where trust is widely discussed and measured [113]. These areas include physiology, sociology and computer science, just to name a few. There are many definitions for trust and some examples are discussed below. Commonly, the entity who trusts another entity is known as the *trustor* and the other entity that is being trusted by the trustor is known as the *trustee*. The mathematical or logical relation between these two entities (i.e. trustor and the trustee) is known as the trust relationship. The trust relationship is measured by the trust assessment. The trust assessment is the process that is used to measure a trust value within a specific context and the period. Typically, trust can be observed as a metric that is gathered by the interactions and observations based on the actors involved in a system.

There is no universally accepted definition for trust [114]. In different disciplines trust can be referred as to the honesty, truthfulness or even the reliability of a trustee. It always varies in different context and the way it is used. Next, we provide a short discussion of trust in two specific areas.

- **Trust in Social Sciences:** Trust in social science integrates the idea of social influence. This comes from characteristics and the behaviour of an individual and can be measured as the honesty, cooperativeness and their willingness to offer help within a social setup. There are wider definitions for describing trust in social sciences. According to [115], *"in psychology, trust is believing that the person who is trusted will do what is expected"*. It starts at the family and grows to others. This indicates the relationship between the persons in a person's life. This defines the aspect of trust from an internal phenomenon (of a person) that helps to maintain a normal relationship between individuals. Rousseau et al. [116] discuss trust as *"a psychological state comprising the intention to accept vulnerability based upon positive expectations of the intentions or behaviour of another"*. Deutsch [117] defines trust as follows: *"an individual may be said to have trust in the occurrence of an event if he expects its occurrence and his expectations to lead to behaviours which he perceives to have greater negative consequences if the expectation is not confirmed than positive motivational consequences"*. This measures the physiological state of a trustor within a specific context and time. The definition provides specific context of the human mind and their functions. The trust relationship in these cases are measured by the behaviour of the actors in a given context.
In sociology, trust is evaluated by the relationship between the social actors. Mcknight and Chervany [38] define trust as the *"extent to which one party is willing to depend on something or somebody in a given situation with a feeling of relative security, even though negative consequences are possible"*. This definition indicates the nature of diversity in the involved social relationship among the entities. Jøsang and Stéphane [51] define trust as *"the extent to which one party is willing to depend on somebody, or something, in a given situation with a feeling of relative security, even though negative consequences are possible"*. This definition indicates the subjective nature of the trust that is dependent upon the willingness and relative security. In a social domain, Lewis and Weigert [118] describe trust as the follows: *"from a sociological perspective, trust must be conceived as a property of collective units (ongoing dyads, groups, and collectivities), not of isolated individuals. Being a collective attribute, trust is applicable to the relations among people rather than to their psychological states taken individually"*. This emphasizes the presence of each entity or their symbolic representations for trust value measurement.
- **Trust in Computing Systems:** In the area of computing and information technology, the notion of trust is used in different areas including networking, security, artificial intelligence, human computer interaction, e-commerce, just to name a few [112]. In computer systems trust can be represented based on subjective belief [119]. Instead of depending upon the context for a specific trust value, some proposals discuss the concept of trust independently. The core concept of trust in computing systems is derived from the social sciences. In general computing systems, trust can be decomposed into various aspects e.g. device trust, entity trust and data trust [120]. Cho et al. [121] define trust as a subjective belief, as follows: *"an agent's trust is a subjective belief about whether another entity will exhibit behaviour reliably in a particular context with potential risks. The agent can make*

a decision based on learning from past experience to maximize its interest (or utility) and/or minimize risk".

Similar to the concept of [121], Jøsang et al. [122] define trust *"as the subjective probability by which an individual expects that another performs a given action on which its welfare depends"*. Mui et al. [123] also define trust as *"a subjective expectation an agent has about another's future behaviour based on the history of their encounters"*. This is to note that, the formal definitions of trust in social sciences can also be extended to computer science. Kimery et al. [124] discuss trust for online systems, in this trust is defined as *"a consumer's willingness to accept vulnerability in an online transaction based on their positive expectations regarding an e-retailer's future behaviors"*. This definition denotes the predicted behaviours of the users in an online system. Cynthia et al. [125] define trust as *"an attitude of confident expectation in an online situation of risk that one's vulnerabilities will not be exploited"*. In this case, the authors consider research on trust between users and online systems (e.g. transaction websites).

Fundamentally, in a computer system, to establish trust is a challenging issue due to the requirements of the specific rules and security policies that are governed by the trustor. If a process stratifies the basic security requirements of a system then the process can be called a 'trusted process'. These policy based mechanisms allow an individual to build trust through the exchange of credentials e.g. digital signatures [126]. The other way of building trust is reputation. Artz and Gil [127] define reputation as: *"reputation is an assessment based on the history of interactions with or observations of an entity, either directly with the evaluator (personal experience) or as reported by others (recommendations or third party verification)"*.

With the introduction of the concept of distributed computing systems and services, the notion of trust becomes a crucial issue due to the pervasive nature of the environments [128]. In such distributed sysytems, the trusted authorities were introduced to manage the trust issues. This is further challenging as the concept of user authentication is not limited to a single domain anymore. Service are available at anytime and anywhere and users require a trusted platform. The Trusted Computing Group (TCG), was formed in 2003, to develop standards and promote security specifications for computers and networks. The TCG introduced Trusted Platform Module (TPM), a hardware chip (microcontroller) integrated with cryptographic keys. It also assess the trustworthiness of a computing system by securely storing artifacts employed to authenticate the platform [129].

2.6.2 Trust Modeling

In this section, we outline the process of trust modelling trust in general computing systems [130]. Trust modeling can be seen as the specific mathematical techniques that are used for modeling trust. In other words, trust modelling comprises a set of rules and languages that are required to establish trust among entities in an automatic

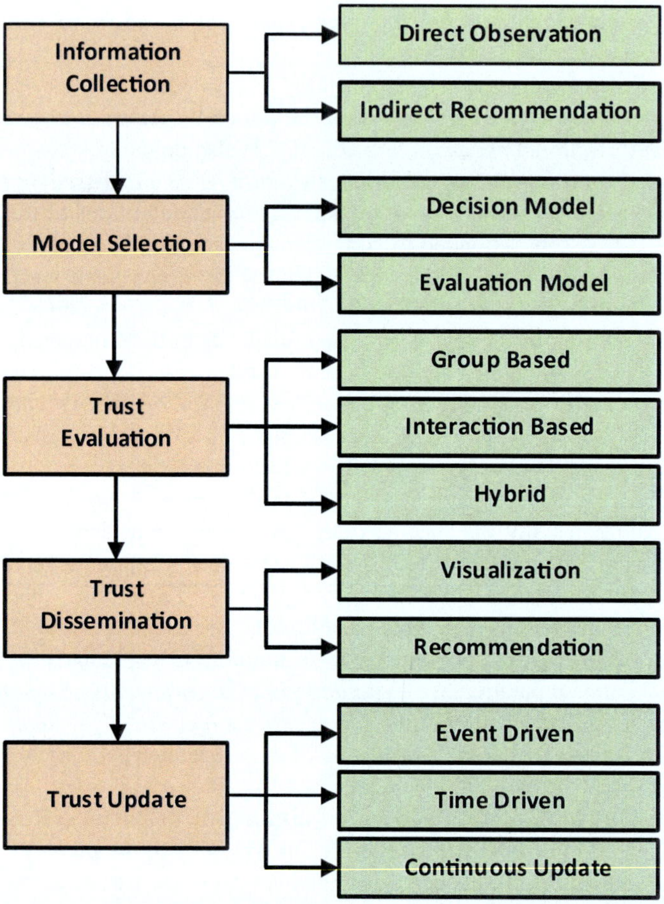

Fig. 2.10 The trust modeling process

or semi-automatic way [131]. Note, when we discuss trust modelling, we consider trust issues both in the social sciences and computing systems. This is due to the interactions between the human users and computing systems which are sometimes essential in many circumstances, or even sometimes unavoidable. For instance, communication between the servers and the human users for some specific application.

In Fig. 2.10, we illustrate a general trust modelling process [132, 133]. It is composed of the following five steps, they are: (i) information collection, (ii) model section, (iii) trust evaluation, (iv) trust dissemination and (v) trust update. We briefly outline these steps as follows:

- In the first step, information is gathered from the nodes (i.e. information is collected by the trustor about the trustee). This can be done in two different ways i.e. direct observation or indirect recommendation [134].

- In the second step, appropriate trust models are used to process the gathered data [135]. Two widely used trust models for this purpose are, decision model and evaluation model. Decision models are typically a policy based approach that restricts the access to resources by means of certain policies [136]. The evaluation models are based on specific attributes of the specific entities. The intention of these trust models are to evaluate readability (the said attributes) by measuring certain factors that have an influence on trust [137].
- In the third step, graph-based or history-based based models are used to calculate the trust value. Sometimes a hybrid based trust method is also used for trust value evaluation [138].
- In the fourth step, two different processes can be used for trust dissemination. First, recommendation models can be used for further recommending trust to other entities in the system. Second, visualization models are used to visualize the trust values [131].
- In the fifth step, trust values are periodically updated. Different methods can be used e.g. event driven (in this case, trust values are updated as soon as the event occurs), time driven (in this case, trust values are updated periodically for all nodes) and continuous update (in this case, the updating of trust values are done based on the integrity protection and verification) [139].

Commonly, there are four widely used trust models available for computing systems. They are statistical trust, machine learning-based trust modelling, heuristical trust modelling and behaviour-based trust modelling [140, 141]. In statistical trust modeling, mathematical calculations are used for the trust value calculation. There are two popular trust models available in this case, Bayesian model and belief model. For the former, the interactions between the entities are defined as a binary rating [142, 143]. For the later, an entity's belief about any statement is translated into rating which is further transformed into an actual trust. In the machine learning-based trust modeling, various machine learning-based algorithms are used to determine a node's behaviour [144]. Heuristical trust modelling is based on the heuristic classification methods which is simple to understand [145]. Finally, in behaviour-based trust modelling, entities communication frequencies, friendliness and cooperation are used [146].

2.6.3 Mechanisms

Now we briefly introduce two commonly used mechanisms for the trust evaluation process that we employ in this book that are related to our present research. They are *belief theory* and *subjective logic*.

2.6.3.1 Belief Theory

It is also known as Dempster–Shafer Theory (DST) or evidence theory introduced by Dempster in the 1960s [147]. It is a general framework for reasoning of traditional probability theory that allows for priori (incomplete) knowledge i.e. with uncertainty. In other words, it assigns a way to measure reasoning under uncertainty using probability, possibility and plausibility with upper and lower bounds. Later in the 1970s, Dempster enhanced this theory by incorporating the notion of evidence from different sources. The evidence is the claim that supports the actual state of an entity (e.g. today's temperature is below freezing). These evidences are collected (both direct and neighbouring of the system) to form a mathematical reasoning to believe the system. The basic idea of DST is the numerical measures of degree of belief from different entities in uncertainty using subjective probability and then combine these beliefs to construct a single belief. The uncertainty is measured by belief and plausibility (i.e. neither belief nor disbelief) rather than pointing to a single value as in the traditional probability theory. Note that the generalized form of the Bayesian theory of subjective probability can be seen as the belief theory [148].

Definition 2.1 (*Belief and Plausibility*) The amount of evidence in favour of a proposition (a) is denoted as belief (Bel) within a range of 0 and 1, where, 0 represents no evidence and 1 represents uncertainty. Then the relationship between the plausibility (Pl) and belief (Bel) can be denoted as follows:

$$Pl(a) = 1 - Bel(a) \ where \ Bel(a) \leq Pl(a) \tag{2.1}$$

Definition 2.2 (*Dempster's Rule of Combination*) It defines the combination of two or more independent set of belief functions [149]. For instance, assume x_1 and x_2 are the two independent sets of belief functions, then the combination can be defined as follows:

$$x_{1,2}(A) = (x_1 \oplus x_2)(A) = \sum_{B \cap C = A \neq \phi} (x_1(B)x_2(C)) \frac{1}{1-K} \tag{2.2}$$

where, A ,B and C are the sets of belief functions. K is the conflict, where, $K = \sum_{B \cap C = \phi}(x_1(B)x_2(C))$ and $x_{1,2}(\phi) = 0$.

2.6.3.2 Subjective Logic

Jøsang [150] defines subjective logic as follows: *"subjective logic is a calculus for subjective opinions which in turn represent probabilities affected by degrees of uncertainty"*. The subjective logic is one type of probabilistic logic that explicitly considers uncertainty during trust evaluation. The arguments of subjective logic are called opinions. An opinion is denoted as ${}^{A}\omega_X$, where A is the source of the said opinion and X represents the state of the variable in which the particular opinion exists. These subjective opinions help to construct the arguments of the subjective logic, where each opinion represents the degree of uncertainty and analysis situations that are consisted of relatively unreliable sources. Each opinion in subjective logic is equivalent to the binomial opinion of the beta distribution function. These functions are employed to calculate the success rate of an event based on the previous knowledge of interactions. A beta distribution function can be represented by $f(\partial|\alpha, \beta)$ and denoted as follows:

$$f(\partial|\alpha, \beta) = \frac{\tau(\alpha+\beta)}{\tau(\alpha)\tau(\beta)} \partial^{\alpha-1}(1-\partial)^{(\beta-1)} \tag{2.3}$$

where, α and β are the evidence parameters, ∂ is the probability variable, $\alpha > 0$, $\beta > 0$ and $0 \leq \partial \leq 1$, $\partial \neq 1$ if $\alpha < 1$ and $\partial \neq 1$ if $\beta < 1$.

Now we discuss how subjective logic can be used for domains containing uncertainty. Let us consider that p and q represent positive and negative experiences of any past interactions, and it is represented as: $\alpha = p + 1$ and $\beta = q + 1$. Now, using values for p and q in the Eq. 2.3, we can create a mapping between the potability distribution function and the priori experiences.

In subjective logic, the term *opinion* represents a belief. In addition, it includes the concepts of disbelief and uncertainty. Note, in this book, we use the notion of belief, disbelief and uncertainty as follows:

- *Belief (b)*: Represents a specified entity's belief for a certain proposition is true.
- *Disbelief (d)*: Represents a specified entity's belief for a certain proposition is not true.
- *Uncertainty (u)*: Represents a specified entity's uncertainty (i.e. neither belief nor disbelief) for a certain proposition.

The relationship among the parameters belief (b), disbelief (d) and uncertainty (u) can be represented as follows:

$$b + d + u = 1, \ \ where\, b, d, u \in [0, 1] \tag{2.4}$$

If γ is the mapping between the Bayesian and belief theories, then γ can be obtained using the following relationship:

$$\gamma = \begin{cases} b = \frac{p}{p+q+2} \\ d = \frac{q}{p+q+2} \\ u = \frac{2}{p+q+2} \end{cases}$$

2.6.4 Category

Broadly there are two ways to calculate trust, direct observations and indirect recommendations [151, 152]. The concepts of trust and reputation are closely related for any trust management systems. The term reputation signifies the opinion of any trustor towards a trustee. This can be measured by the past experiences and the interactions between the trustor and the trustee.

- *Direct Observations:* In this case, the direct observation of an entity about other entities is taken for calculating a trust value (i.e. trustworthiness of interaction partners is observed). For instance, user Alice wants to access a specific service from a service provider, Bob. Alice can trust Bob based on their previous interactions or other positive experiences. This enhances a direct trust between them as them. The direct trust is typically measured and calculated over the reputation. Note, it is not always necessary that the entities know one another directly or have had previous interactions. This leads to the indirect recommendations discussed below.
- *Indirect Recommendations:* In this case, one entity takes information or trust recommendations of another entity through other entities. For instance, user Alice wants to access a service from a service provider Bob. They do not have any previous interactions. In this situation, Alice can take a recommendation from her friend John who had a positive interaction in the past with Bob. This is a transitive trust calculation process where the trust value computation between Alice and Bob is based on two separate processes i.e. Alice's trust with John and John's trust with Bob.

2.7 Summary

In this chapter, we have discussed background and related work on the state of the art IoT and its access control issue. We argue that enforcing security policies and developing appropriate security requirements for the IoT has not only become an essential issue but also an obligation. Our study indicated the need to protect IoT systems and resources from potential threats and attacks not only in internal networks but also originating from networks that span over multiple domains. With the sensitive nature of the IoT and its dynamic characteristics, many of these issues cannot be addressed with a simple software patch or commonly used security measures. We note that the research in access control for IoT is still in its infancy without enough attention is being paid to access control governed by the security, identity, delegation and trust. Therefore, there is significant need for appropriate access control mechanisms to ensure a secure IoT system. We further argue that the, IoT goes beyond to the Internet and the traditional Internet architecture is not enough to handle the access control mechanisms for the IoT. In such, *things* need to be configured for its operating conditions, criteria, sensitivity that are then used for controlling the entire system operation.

In the next chapter (Chap. 3), we plan to discuss the design and development of an access control architecture for large-scale IoT systems. Our intention is to examine how to manage the vast amount of users, heterogeneous devices, applications and their associated services in an IoT system, especially to different requirements and access control issues (e.g. efficient and fine-grained policy management and their enforcement) that we discussed in this chapter.

References

1. L. Atzori, A. Iera, G. Morabito, The internet of things: a survey. Comput. Netw. **54**(15), 2787–2805 (2010). http://dx.doi.org/10.1016/j.comnet.2010.05.010
2. J. Gubbi, R. Buyya, S. Marusic, M. Palaniswami, Internet of Things (IoT): a vision, architectural elements, and future directions. Future Gener. Comput. Syst. **29**(7), 1645–1660 (2013). http://dx.doi.org/10.1016/j.future.2013.01.010
3. L. Tan, N. Wang, Future internet: the internet of things, in *3rd International Conference on Advanced Computer Theory and Engineering(ICACTE)* (IEEE, 2010), pp. V5–376–V5–380. http://dx.doi.org/10.1109/icacte.2010.5579543
4. S. Haller, S. Karnouskos, C. Schroth, The internet of things in an enterprise context, in *Future Internet*, ed. by J. Domingue, D. Fensel, P. Traverso. Lecture Notes in Computer Science, vol. 5468 (Springer, Berlin, 2009), pp. 14–28. http://dx.doi.org/10.1007/978-3-642-00985-3_2
5. L. Davoli, L. Veltri, G. Ferrari, U. Amadei, Internet of Things on Power Line Communications: An Experimental Performance Analysis (Springer, Singapore, 2019), pp. 465–498. https://doi.org/10.1007/978-981-13-1768-2_13
6. A. Dhumane, R. Prasad, J. Prasad, Routing issues in internet of things: a survey. Proceedings of the international multiconference of engineers and computer scientists **1**, 16–18 (2016)
7. P. Aswale, A. Shukla, P. Bharati, S. Bharambe, S. Palve, *An overview of internet of things: Architecture, protocols and challenges, in Information and Communication Technology for Intelligent Systems* (Springer, Berlin, 2019), pp. 299–308
8. N. Neshenko, E. Bou-Harb, J. Crichigno, G. Kaddoum, N. Ghani, Demystifying iot security: An exhaustive survey on iot vulnerabilities and a first empirical look on internet-scale iot exploitations. IEEE Commun. Surv. Tutor. 1–1 (2019)
9. A. Al-Fuqaha, M. Guizani, M. Mohammadi, M. Aledhari, M. Ayyash, Internet of things: a survey on enabling technologies, protocols, and applications. IEEE Commun. Surv. & Tutor. **17**(4), 2347–2376 (2015). http://dx.doi.org/10.1109/comst.2015.2444095
10. P. Ray, A survey on Internet of Things architectures - ScienceDirect, http://www.sciencedirect.com/science/article/pii/S1319157816300799
11. G. Manogaran, R. Varatharajan, D. Lopez, P.M. Kumar, R. Sundarasekar, C. Thota, A new architecture of internet of things and big data ecosystem for secured smart healthcare monitoring and alerting system. Future Gen. Comput. Syst. **82**, 375–387 (2018), http://www.sciencedirect.com/science/article/pii/S0167739X17305149
12. R. Khan, S.U. Khan, R. Zaheer, S. Khan, Future internet: the internet of things architecture, possible applications and key challenges, in *2012 10th International Conference on Frontiers of Information Technology (FIT)* (IEEE, 2012), pp. 257–260. http://dx.doi.org/10.1109/fit.2012.53
13. M. Wu, T.-J. Lu, F.-Y. Ling, J. Sun, H.-Y. Du, Research on the architecture of Internet of Things, in *2010 3rd International Conference on Advanced Computer Theory and Engineering(ICACTE)* (IEEE, 2010), pp. V5–484–V5–487. http://dx.doi.org/10.1109/icacte.2010.5579493
14. J.E. Siegel, D.C. Erb, S.E. Sarma, A survey of the connected vehicle landscape?architectures, enabling technologies, applications, and development areas. IEEE Trans. Intell. Transp. Syst. **19**(8), 2391–2406 (2018)
15. S. Li, T. Tryfonas, H. Li, The Internet of Things:a security point of view. Int. Res. **26**(2), 337–359 (2016), http://www.emeraldinsight.com/doi/pdfplus/10.1108/IntR-07-2014-0173

16. Cisco: The internet of things reference model, http://cdn.iotwf.com/. Accessed 04 Oct 2018
17. C. Perera, A. Zaslavsky, P. Christen, D. Georgakopoulos, Context aware computing for the internet of things: a survey. IEEE Commun. Surv. & Tutor. **16**(1), 414–454 (2014). http://dx.doi.org/10.1109/surv.2013.042313.00197
18. F. Wang, L. Hu, J. Hu, J. Zhou, K. Zhao, Recent advances in the internet of things: multiple perspectives. IETE Techn. Rev., 1–11 (2016). http://dx.doi.org/10.1080/02564602.2016.1155419
19. F. Javed, M.K. Afzal, M. Sharif, B. Kim, Internet of things (iot) operating systems support, networking technologies, applications, and challenges: a comparative review. IEEE Commun. Surv. Tutor. **20**(3), 2062–2100, thirdquarter (2018)
20. L.D. Xu, W. He, S. Li, Internet of things in industries: a survey. IEEE Trans. Ind. Inf. **10**(4), 2233–2243 (2014)
21. A.A. Zaidan, B.B. Zaidan, A review on intelligent process for smart home applications based on iot: coherent taxonomy, motivation, open challenges, and recommendations. Artif. Intell. Rev. (2018). https://doi.org/10.1007/s10462-018-9648-9
22. B.L.R. Stojkoska, K.V. Trivodaliev, A review of internet of things for smart home: Challenges and solutions. J. Cleaner Prod. **140**, 1454–1464 (2017), http://www.sciencedirect.com/science/article/pii/S095965261631589X
23. M.B. Yassein, M.Q. Shatnawi, D. Al-zoubi, Application layer protocols for the internet of things: a survey, in *2016 International Conference on Engineering MIS (ICEMIS)* (2016), pp. 1–4
24. S.N. Swamy, D. Jadhav, N. Kulkarni, Security threats in the application layer in iot applications, in *2017 International Conference on I-SMAC (IoT in Social, Mobile, Analytics and Cloud) (I-SMAC)* (2017), pp. 477–480
25. V. Karagiannis, P. Chatzimisios, F. Vazquez-Gallego, J. Alonso-Zarate, A survey on application layer protocols for the internet of things. Trans. IoT Cloud Comput. **3**(1), 11–17 (2015)
26. A. Riahi, E. Natalizio, Y. Challal, N. Mitton, A. Iera, A systemic and cognitive approach for iot security, in *2014 International Conference on Computing, Networking and Communications (ICNC)* (2014), pp. 183–188. http://dx.doi.org/10.1109/ICCNC.2014.6785328
27. M. Ammar, G. Russello, B. Crispo, Internet of things: a survey on the security of iot frameworks. J. Inf. Sec. Appl. **38**, 8–27 (2018), http://www.sciencedirect.com/science/article/pii/S2214212617302934
28. S. Pattar, R. Buyya, K.R. Venugopal, S.S. Iyengar, L.M. Patnaik, Searching for the iot resources: Fundamentals, requirements, comprehensive review, and future directions. IEEE Commun. Surv. Tutor. **20**(3), 2101–2132, thirdquarter (2018). http://dx.doi.org/10.1109/COMST.2018.2825231
29. H. Cai, B. Xu, L. Jiang, A.V. Vasilakos, Iot-based big data storage systems in cloud computing: perspectives and challenges. IEEE Int. Things J. **4**(1), 75–87 (2017). http://dx.doi.org/10.1109/JIOT.2016.2619369
30. C. Stergiou, K.E. Psannis, B.-G. Kim, B. Gupta, Secure integration of iot and cloud computing. Future Gen. Comput. Syst. **78**, 964 – 975 (2018), http://www.sciencedirect.com/science/article/pii/S0167739X1630694X
31. M. Zorzi, A. Gluhak, S. Lange, A. Bassi, From today's intranet of things to a future internet of things: a wireless- and mobility-related view. IEEE Wireless Commun. **17**(6), 44–51 (2010). http://dx.doi.org/10.1109/MWC.2010.5675777
32. E. Siow, T. Tiropanis, W. Hall, Analytics for the internet of things: a survey. ACM Comput. Surv. **51**(4), 74:1–74:36 (2018). http://doi.acm.org/10.1145/3204947
33. A. Tchernykh, U. Schwiegelsohn, E. ghazali Talbi, M. Babenko, Towards understanding uncertainty in cloud computing with risks of confidentiality, integrity, and availability. J. Comput. Sci. (2016), http://www.sciencedirect.com/science/article/pii/S1877750316303878
34. Access control definition, https://dictionary.cambridge.org/dictionary/english/access-control. Accessed 02 Jun 2018
35. R.S. Sandhu, P. Samarati, Access control: principle and practice. IEEE Commun. Mag. **32**(9), 40–48 (1994)

36. R. Sandhu, P. Samarati, Authentication, access control, and audit. ACM Comput. Surv. **28**(1), 241–243 (1996). http://doi.acm.org/10.1145/234313.234412
37. h. . u. y. . . n. . O. a. .-J.-. Wikibooks, title = Fundamentals of Information Systems Security/Access Control Systems
38. P. Samarati, S.C. de Vimercati, Access control: policies, models, and mechanisms, in *Foundations of Security Analysis and Design*, ed. by R. Focardi, R. Gorrieri (Springer, Berlin, 2001), pp. 137–196
39. S. Gusmeroli, S. Piccione, D. Rotondi, A capability-based security approach to manage access control in the Internet of Things. Math. Comput. Model. **58**(5–6), 1189–1205 (2013). http://dx.doi.org/10.1016/j.mcm.2013.02.006
40. F. Cai, N. Zhu, J. He, P. Mu, W. Li, Y. Yu, Survey of access control models and technologies for cloud computing (2018), pp. 1–12. http://dx.doi.org/10.1007/s10586-018-1850-7
41. A. Alshehri, R. Sandhu, Access control models for cloud-enabled internet of things: a proposed architecture and research agenda, in *2016 IEEE 2nd International Conference on Collaboration and Internet Computing (CIC)* (2016), pp. 530–538
42. N. Fotiou, T. Kotsonis, G.F. Marias, G.C. Polyzos, Accesscontrol forthe internet of things, in *International Workshop on SecureInternet of Things (SIoT)* (2016), pp. 29–38
43. L. Xu, H. Zhang, X. Du, C. Wang, Research on mandatory access control model for application system, in *2009 International Conference on Networks Security, Wireless Communications and Trusted Computing*, vol. 2 (2009), pp. 159–163
44. J.A. Solworth, R.H. Sloan, A layered design of discretionary access controls with decidable safety properties, in *IEEE Symposium on Security and Privacy, 2004. Proceedings* (2004), pp. 56–67
45. D.F. Ferraiolo, R. Sandhu, S. Gavrila, D.R. Kuhn, R. Chandramouli, Proposed NIST standard for role-based access control. ACM Trans. Inf. Syst. Sec. **4**(3), 224–274 (2001). http://dx.doi.org/10.1145/501978.501980
46. E. Yuan, J. Tong, Attributed Based Access Control (ABAC) for web services, in *Proceedings of the IEEE International Conference on Web Services*, ser. ICWS '05 (IEEE Computer Society, Washington, DC, USA, 2005), pp. 561–569. http://dx.doi.org/10.1109/icws.2005.25
47. L. Gong, A secure identity-based capability system, in *Proceedings of the 1989 IEEE Symposium on Security and Privacy* (1989), pp. 56–63, http://citeseerx.ist.psu.edu/viewdoc/summary?doi=10.1.1.53.1785
48. T. Kothmayr, C. Schmitt, W. Hu, M. Brünig, G. Carle, Dtls based security and two-way authentication for the internet of things. Ad Hoc Netw. **11**(8), 2710–2723 (2013)
49. D. Hardt, The oauth 2.0 authorization framework, RFC, internet engineering task force (ietf). Technical Report (2012), http://www.rfc-editor.org/info/rfc6749
50. V. Suhendra, A survey on access control deployment, in *Security Technology*, ed. by T.-H. Kim, H. Adeli, W.-C. Fang, J.G. Villalba, K.P. Arnett, M.K. Khan (Springer, Berlin, 2011), pp. 11–20
51. S. De Capitani, P. di Vimercati, S.Jajodia Samarati, Policies, models, and languages for access control, in *Databases in Networked Information Systems*, ed. by S. Bhalla (Springer, Berlin, 2005), pp. 225–237
52. A. Vakali, *Access Control Policy Languages* (Springer, Boston, 2009), pp. 15–18. https://doi.org/10.1007/978-0-387-39940-9_5
53. X. Wang, G. Lao, T. DeMartini, H. Reddy, M. Nguyen, E. Valenzuela, Xrml – extensible rights markup language, in *Proceedings of the 2002 ACM Workshop on XML Security*, ser. XMLSEC '02 (ACM, New York, NY, USA, 2002), pp. 71–79. http://doi.acm.org/10.1145/764792.764803
54. extensible access control markup language (xacml) version 3.0 (2013), http://docs.oasis-open.org/xacml/3.0/xacml-3.0-core-spec-os-en.html. Accessed 02 Nov 2017
55. D. Ferraiolo, R. Chandramouli, R. Kuhn, V. Hu, Extensible access control markup language (xacml) and next generation access control (ngac), in *Proceedings of the 2016 ACM International Workshop on Attribute Based Access Control* (ACM, 2016), pp. 13–24. https://doi.org/10.1145/2875491.2875496

56. H. Jiang, A. Bouabdallah, Jacpol: a simple but expressive json-based access control policy language, in *Information Security Theory and Practice*, ed. by G.P. Hancke, E. Damiani (Springer International Publishing, Cham, 2018), pp. 56–72
57. J. Crampton, C. Morisset, Ptacl: a language for attribute-based access control in open systems, in *Principles of Security and Trust*, ed. by P. Degano, J.D. Guttman (Springer, Berlin, 2012), pp. 390–409
58. R. Mathur, S. Agarwal, V. Sharma, Solving security issues in mobile computing using cryptography techniques–a survey, in *International Conference on Computing, Communication Automation* (2015), pp. 492–497
59. A.R. Parab, Generic approach for encryption using reverse context free grammar productions. Int. J. Res. Appl. Sci. Eng. Technol. **6**, 197–202, 03 (2018)
60. S. Chandra, S. Paira, S.S. Alam, G. Sanyal, A comparative survey of symmetric and asymmetric key cryptography, in *2014 International Conference on Electronics, Communication and Computational Engineering (ICECCE)* (2014), pp. 83–93
61. A. Setiawan, D. Adiutama, J. Liman, A. Luther, R. Buyya, Gridcrypt: high performance symmetric key cryptography using enterprise grids, in *Parallel and Distributed Computing: Applications and Technologies*, ed. by K.-M. Liew, H. Shen, S. See, W. Cai, P. Fan, S. Horiguchi (Springer, Berlin, 2005), pp. 872–877
62. B. Schneier, Description of a new variable-length key, 64-bit block cipher (blowfish), in *Fast Software Encryption*, ed. by R. Anderson (Springer, Berlin, 1994), pp. 191–204
63. A.K. Mandal, C. Parakash, A. Tiwari, Performance evaluation of cryptographic algorithms: Des and aes, in *2012 IEEE Students' Conference on Electrical, Electronics and Computer Science* (2012), pp. 1–5
64. L. Batina, J. Guajardo, T. Kerins, N. Mentens, P. Tuyls, I. Verbauwhede, Public-key cryptography for rfid-tags, in *Fifth Annual IEEE International Conference on Pervasive Computing and Communications Workshops (PerComW'07)* (2007), pp. 217–222
65. P.C. van Oorschot, M.J. Wiener, On diffie-hellman key agreement with short exponents, in *Advances in Cryptology – EUROCRYPT '96*, ed. by U. Maurer (Springer, Berlin, 1996), pp. 332–343
66. P. MacKenzie, S. Patel, R. Swaminathan, Password-authenticated key exchange based on rsa, in *Advances in Cryptology - ASIACRYPT*, ed. by T. Okamoto (Springer, Berlin, 2000), pp. 599–613
67. N. Koblitz, A. Menezes, S. Vanstone, The state of elliptic curve cryptography. Des. Codes Cryptograp. **19**(2), 173–193 (2000). https://doi.org/10.1023/A:1008354106356
68. L. Da Xu, W. He, S. Li, Internet of things in industries: a survey. IEEE Trans. Ind. Inf. **10**(4), 2233–2243 (2014)
69. J. Hernandez-Ramos, A. Jara, L. Marın, A. Skarmeta, Distributed capability-based access control for the internet of things. J. Int. Serv. Inf. Sec. **3**(3/4), 1–16 (2013), http://isyou.info/jisis/vol3/no34/jisis-2013-vol3-no34-01.pdf
70. Y. Lee, J. Lim, Y. Jeon, J. Kim, Technology trends of access control in iot and requirements analysis, in *2015 International Conference on Information and Communication Technology Convergence (ICTC)* (2015), pp. 1031–1033
71. D. Hussein, E. Bertin, V. Frey, Access control in iot: From requirements to a candidate vision, in *2017 20th Conference on Innovations in Clouds, Internet and Networks (ICIN)* (2017), pp. 328–330
72. M.P. Michael, M. Darianian, Architectural solutions for mobile rfid services for the internet of things, in *2008 IEEE Congress on Services - Part I* (2008), pp. 71–74
73. D. Kulkarni, A. Tripathi, Context-aware role-based access control in pervasive computing systems, in *Proceedings of the 13th ACM Symposium on Access Control Models and Technologies*, ser. SACMAT '08 (ACM, New York, NY, USA, 2008), pp. 113–122. http://doi.acm.org/10.1145/1377836.1377854
74. G. Zhang, J. Tian, An extended role based access control model for the internet of things, in *2010 International Conference on Information, Networking and Automation (ICINA)*, vol. 1 (2010), pp. V1-319–V1-323

75. G. Zhang, M. Parashar, Context-aware dynamic access control for pervasive applications, in *Proceedings of the Communication Networks and Distributed Systems Modeling and Simulation Conference* (2004), pp. 21–30
76. A.A. El Kalam, S. Benferhat, R. El Baida, C. Saurel, P. Balbiani, Y. Deswarte, G. Trouessin, et al., Organization based access control, in *null* (IEEE, 2003), p. 120
77. I.B. Pasquier, A.A. Ouahman, A.A.E. Kalam, M.O. de Montfort, Smartorbac security and privacy in the internet of things, in *2015 IEEE/ACS 12th International Conference of Computer Systems and Applications (AICCSA)* (2015), pp. 1–8
78. E. Freudenthal, T. Pesin, L. Port, E. Keenan, V. Karamcheti, drbac: distributed role-based access control for dynamic coalition environments, in *Proceedings 22nd International Conference on Distributed Computing Systems* (2002), pp. 411–420. http://doi.acm.org/10.1109/ICDCS.2002.1022279
79. M. Burnside, D. Clarke, S. Devadas, R. Rivest, Distributed SPKI/SDSI-Based Security for Networks of Devices (CSG, 2002), http://citeseerx.ist.psu.edu/viewdoc/summary?doi=10.1.1.550.1516
80. J. Liu, Y. Xiao, C.L.P. Chen, Authentication and access control in the internet of things, in *2012 32nd International Conference on Distributed Computing Systems Workshops* (2012), pp. 588–592
81. Q. Liu, H. Zhang, J. Wan, X. Chen, An access control model for resource sharing based on the role-based access control intended for multi-domain manufacturing internet of things. IEEE Access **5**, 7001–7011 (2017)
82. J. Jindou, Q. Xiaofeng, C. Cheng, Access control method for web of things based on role and sns, in *2012 IEEE 12th International Conference on Computer and Information Technology* (2012), pp. 316–321
83. E. Barka, S.S. Mathew, Y. Atif, Securing the web of things with role-based access control, in *Codes, Cryptology, and Information Security*, ed. by S. El Hajji, A. Nitaj, C. Carlet, E.M. Souidi (Springer International Publishing, Cham, 2015), pp. 14–26
84. D. Guinard, V. Trifa, *Towards the web of things: web mashups for embedded devices, in Workshop on Mashups, Enterprise Mashups and Lightweight Composition on the Web (MEM 2009), in proceedings of WWW (International World Wide Web Conferences)*, vol. 15 (Madrid, Spain, 2009)
85. G. Zhang, J. Liu, A model of workflow-oriented attributed based access control. Int. J. Comput. Netw. Inf. Sec. **3**(1), 47 (2011)
86. B. Bezawada, K. Haefner, I. Ray, Securing home iot environments with attribute-based access control, in *Proceedings of the Third ACM Workshop on Attribute-Based Access Control*, ser. ABAC'18 (ACM, New York, NY, USA, 2018), pp. 43–53. http://doi.acm.org/10.1145/3180457.3180464
87. N. Ye, Y. Zhu, R.-c. Wang, Q.-m. Lin, An efficient authentication and access control scheme for perception layer of internet of things. (Natural Sciences Publishing Cor., 2014), https://repository.up.ac.za/handle/2263/39762
88. L. Touati, Y. Challal, Poster: activity-based access control for iot, in *Proceedings of the 1st International Workshop on Experiences with the Design and Implementation of Smart Objects* (ACM, 2015), pp. 29–30
89. S. Sciancalepore, M. Pilc, S. Schröder, G. Bianchi, G. Boggia, M. Pawłowski, G. Piro, M. Płóciennik, H. Weisgrab, Attribute-based access control scheme, in *federated iot platforms, in Interoperability and Open-Source Solutions for the Internet of Things*, ed. by I. Podnar Žarko, A. Broering, S. Soursos, M. Serrano (Springer International Publishing, Cham, 2017), pp. 123–138
90. U. Lang, R. Schreiner, Proximity-based access control (pbac) using model-driven security, in *ISSE 2015*, ed. by H. Reimer, N. Pohlmann, W. Schneider (Springer Fachmedien Wiesbaden, Wiesbaden, 2015), pp. 157–170
91. U. Lang, R. Schreiner, Model driven security (mds) management & enforcement to support soa-style agility, in *Proceedings of ISSE* (2007)

92. J. Park, R. Sandhu, Towards usage control models: beyond traditional access control, in *Proceedings of the seventh ACM symposium on Access control models and technologies* (ACM, 2002), pp. 57–64
93. J. Park, R. Sandhu, The uconabc usage control model. ACM Trans. Inf. Syst. Secur. **7**(1), 128–174 (2004). http://doi.acm.org/10.1145/984334.984339
94. Z. Guoping, G. Wentao, The research of access control based on ucon in the internet of things. J. Softw. **6**(4), 724–731 (2011)
95. S. Gusmeroli, S. Piccione, D. Rotondi, Iot access control issues: a capability based approach, in *2012 Sixth International Conference on Innovative Mobile and Internet Services in Ubiquitous Computing* (2012), pp. 787–792
96. A.F. Skarmeta, J.L. Hernández-Ramos, M.V. Moreno, A decentralized approach for security and privacy challenges in the internet of things, in *2014 IEEE World Forum on Internet of Things (WF-IoT)* (2014), pp. 67–72
97. J.L. Hernndez-Ramos, M.P. Pawlowski, A.J. Jara, A.F. Skarmeta, L. Ladid, Toward a lightweight authentication and authorization framework for smart objects. IEEE J. Select. Areas Commun. **33**(4), 690–702 (2015)
98. A. Bassi, M. Bauer, M. Fiedler, R. Kranenburg, *Enabling Things to Talk* (Springer, Berlin, 2013)
99. Q. Zhou, M. Elbadry, F. Ye, Y. Yang, Flexible, fine grained access control for internet of things: poster abstract, in *Proceedings of the Second International Conference on Internet-of-Things Design and Implementation*, ser. IoTDI '17 (ACM, New York, 2017), pp. 333–334, http://doi.acm.org/10.1145/3054977.3057308
100. Q. Zhou, M. Elbadry, F. Ye, Y. Yang, Heracles: scalable, fine-grained access control forinternet-of-things in enterprise environments, in *IEEE INFOCOM 2018 -IEEE Conference on Computer Communications* (2018), pp. 1772–1780. http://doi.acm.org/10.1109/INFOCOM.2018.8485944
101. B. Anggorojati, P.N. Mahalle, N.R. Prasad, R. Prasad, Capability-based access control delegation model on the federated iot network, in *The 15th International Symposium on Wireless Personal Multimedia Communications* (2012), pp. 604–608
102. R. Xu, Y. Chen, E. Blasch, G. Chen, A federated capability-based access control mechanism for internet of things (iots), in *Sensors and Systems for Space Applications XI*, vol. 10641 (International Society for Optics and Photonics, 2018), p. 106410U
103. P. Mahalle, B. Anggorojati, N.R. Prasad, R. Prasad, Identity establishment and capability based access control (iecac) scheme for internet of things, in *WPMC* (2012), pp. 187–191
104. P.y. Butkus, Identity management in m2m networks, master thesis, aalto university
105. M.A. Bishop, *Computer Security: Art and Science* (Pearson Education, USA, 2002)
106. J. Chen, Y. Liu, Y. Chai, An identity management framework for internet of things, in *2015 IEEE 12th International Conference on e-Business Engineering* (IEEE, 2015), pp. 360–364. http://dx.doi.org/10.1109/icebe.2015.67
107. Y. Cao, L. Yang, A survey of Identity Management technology, in *2010 IEEE International Conference on Information Theory and Information Security* (IEEE, 2010), pp. 287–293. http://dx.doi.org/10.1109/icitis.2010.5689468
108. A. Jøsang, S. Pope, User Centric Identity Management (2005), https://link.springer.com/content/pdf/10.1007%2F978-3-642-23300-5_1.pdf
109. Wikipedia, "Delegation," https://en.wikipedia.org/wiki/Delegation. Accessed 25 May 2019
110. B. Anggorojati, P. N. Mahalle, N.R. Prasad, R. Prasad, Capability-based access control delegation model on the federated IoT network, in *2012 15th International Symposium on Wireless Personal Multimedia Communications (WPMC)* (IEEE, 2012), pp. 604–608, http://ieeexplore.ieee.org/xpls/abs_all.jsp?arnumber=6398784
111. H. Shen, A capability-based access control framework with delegation support, in *Wireless Communications, Networking and Applications*, ed. by Q. Zeng. Lecture Notes in Electrical Engineering, vol. 348 (Springer, India, 2016), pp. 655–667. http://dx.doi.org/10.1007/978-81-322-2580-5_59

112. C.D. Jensen, The importance of trust in computer security, in *Trust Management VIII*, ed. by J. Zhou, N. Gal-Oz, J. Zhang, E. Gudes (Springer, Berlin, 2014), pp. 1–12
113. J.-H. Cho, K. Chan, S. Adali, A survey on trust modeling. ACM Comput. Surv. **48**(2), 28:1–28:40 (2015). http://doi.acm.org/10.1145/2815595
114. K. Arai, Defining trust using expected utility theory. Hitotsubashi J. Econ. 205–224 (2009)
115. Trust (social science) - wikipedia, https://en.wikipedia.org/wiki/Trust_(social_science)
116. D.M. Rousseau, S.B. Sitkin, R.S. Burt, C. Camerer, Not so different after all: a cross-discipline view of trust. Acad. Manag. Rev. **23**(3), 393–404 (1998)
117. M. Deutsch, Cooperation and trust: some theoretical notes (1962)
118. J.D. Lewis, A. Weigert, Trust as a social reality. Social forces **63**(4), 967–985 (1985)
119. A. Jøsang, R. Hayward, S. Pope, Trust network analysis with subjective logic, in *Proceedings of the 29th Australasian Computer Science Conference - Volume 48*, ser. ACSC '06 (Australian Computer Society, Inc., Darlinghurst, 2006), pp. 85–94, http://dl.acm.org/citation.cfm?id=1151699.1151710
120. A. Arabsorkhi, M. Sayad Haghighi, R. Ghorbanloo, A conceptual trust model for the internet of things interactions, in *2016 8th International Symposium on Telecommunications (IST)* (2016), pp. 89–93
121. J. Cho, A. Swami, I. Chen, A survey on trust management for mobile ad hoc networks. IEEE Commun. Surv. Tutor. **13**(4), 562–583 (2011)
122. A. Jsang, R. Ismail, C. Boyd, A survey of trust and reputation systems for online service provision. Decis. Support Syst. **43**(2), 618–644 (2007), Emerging Issues in Collaborative Commerce, http://www.sciencedirect.com/science/article/pii/S0167923605000849
123. L. Mui, Computational models of trust and reputation: agents, evolutionary games, and social networks. Ph.D. dissertation, Massachusetts Institute of Technology (2002)
124. K.M. Kimery, M. McCord, Third party assurances: mapping the road to trust in eretailing. J. Inf. Technol. Theory Appl. (JITTA) **4**(2), 7 (2002)
125. C.L. Corritore, B. Kracher, S. Wiedenbeck, On-line trust: concepts, evolving themes, a model. Int. J. Human-Comput. Stud. **58**(6), 737–758 (2003), Trust and Technology, http://www.sciencedirect.com/science/article/pii/S1071581903000417
126. V. Gligor, J.M. Wing, Towards a theory of trust in networks of humans and computers, in *Security Protocols XIX*, ed. by B. Christianson, B. Crispo, J. Malcolm, F. Stajano (Springer, Berlin, 2011), pp. 223–242
127. D. Artz, Y. Gil, A survey of trust in computer science and the semantic web. J. Web Semant. **5**(2), 58–71 (2007), Software Engineering and the Semantic Web, http://www.sciencedirect.com/science/article/pii/S1570826807000133
128. D. Xiu, Z. Liu, A formal definition for trust in distributed systems, in *Information Security*, ed. by J. Zhou, J. Lopez, R.H. Deng, F. Bao (Springer, Berlin, 2005), pp. 482–489
129. B. Parno, Bootstrapping trust in a "trusted" platform, in *HotSec* (2008)
130. J. Guo, I.-R. Chen, J.J. Tsai, A survey of trust computation models for service management in internet of things systems. Comput. Commun. **97**, 1–14 (2017), http://www.sciencedirect.com/science/article/pii/S0140366416304959
131. C. Fernandez-Gago, F. Moyano, J. Lopez, Modelling trust dynamics in the internet of things. Inf. Sci. **396**, 72–82 (2017), http://www.sciencedirect.com/science/article/pii/S0020025517305364
132. A. Altaf, H. Abbas, F. Iqbal, A. Derhab, Trust models of internet of smart things: a survey, open issues, and future directions. J. Netw. Comput. Appl. **137**, 93–111 (2019), http://www.sciencedirect.com/science/article/pii/S1084804519300839
133. F. Moyano, C. Fernandez-Gago, J. Lopez, A conceptual framework for trust models, in *Trust, Privacy and Security in Digital Business*, ed. by S. Fischer-Hübner, S. Katsikas, G. Quirchmayr (Springer, Berlin, 2012), pp. 93–104
134. Q. Nguyen Vu, S. Hassas, F. Armetta, B. Gaudou, R. Canal, Combining trust and self-organization for robust maintaining of information coherence in disturbed mas, in *2011 IEEE Fifth International Conference on Self-Adaptive and Self-Organizing Systems* (2011), pp. 178–187

135. I. Chen, F. Bao, J. Guo, Trust-based service management for social internet of things systems. IEEE Trans. Dependable Sec. Comput. **13**(6), 684–696 (2016)
136. M. Blaze, J. Feigenbaum, J. Lacy, Decentralized trust management, in *Proceedings 1996 IEEE Symposium on Security and Privacy* (1996), pp. 164–173
137. S.P. Marsh, Formalising trust as a computational concept (1994)
138. M. Li, X. Sun, H. Wang, Y. Zhang, J. Zhang, Privacy-aware access control with trust management in web service. World Wide Web **14**(4), 407–430 (2011). https://doi.org/10.1007/s11280-011-0114-8
139. A. Sharma, E.S. Pilli, A.P. Mazumdar, M.C. Govil, A framework to manage trust in internet of things, in *2016 International Conference on Emerging Trends in Communication Technologies (ETCT)* (2016), pp. 1–5
140. A. Manna, A. Sengupta, C. Mazumdar, A survey of trust models for enterprise information systems. Procedia Comput. Sci. **85**, 527–534 (2016), International Conference on Computational Modelling and Security (CMS, 2016), http://www.sciencedirect.com/science/article/pii/S1877050916305609
141. I. Ud Din, M. Guizani, B. Kim, S. Hassan, M. Khurram Khan, Trust management techniques for the internet of things: a survey. IEEE Access **7**, 29 763–29 787 (2019)
142. B. Liu, A survey on trust modeling from a bayesian perspective (2018). arXiv:1806.03916
143. A. Jøsang, Artificial reasoning with subjective logic, in *Proceedings of the second Australian workshop on commonsense reasoning*, vol. 48 (Citeseer, 1997), p. 34
144. K. Zhao, L. Pan, A machine learning based trust evaluation framework for online social networks, in *2014 IEEE 13th International Conference on Trust, Security and Privacy in Computing and Communications* (2014), pp. 69–74
145. K. Chen, K. Hwang, G. Chen, Heuristic discovery of role-based trust chains in peer-to-peer networks. IEEE Trans. Parallel Distrib. Syst. **20**(1), 83–96 (2009)
146. Z. Yan, V. Niemi, Y. Dong, G. Yu, A user behavior based trust model for mobile applications, in *Autonomic and Trusted Computing*, ed. by C. Rong, M.G. Jaatun, F.E. Sandnes, L.T. Yang, J. Ma (Springer, Berlin, 2008), pp. 455–469
147. A.P. Dempster, Upper and lower probabilities induced by a multivalued mapping. Ann. Math. Statist. **38**, 325–339 (1967)
148. M. Beynon, B. Curry, P. Morgan, The dempster–shafer theory of evidence: an alternative approach to multicriteria decision modelling. Omega **28**(1), 37–50 (2000), http://www.sciencedirect.com/science/article/pii/S030504839900033X
149. A. Josang, S. Pope, Dempster's rule as seen by little colored balls. Comput. Intell. **28**(4), 453–474 (2012). http://dx.doi.org/10.1111/j.1467-8640.2012.00421.x
150. A. Jøsang, A logic for uncertain probabilities. Int. J. Uncertain. Fuzziness Knowl.-Based Syst. **9**(3), 279–311 (2001). http://dx.doi.org/10.1142/S0218488501000831
151. S.K. Prajapati, S. Changder, A. Sarkar, Trust management model for cloud computing environment. vol. abs/1304.5313 (2013). arXiv:1304.5313
152. A.B. Usman, J. Gutierrez, Toward trust based protocols in a pervasive and mobile computing environment: a survey. Ad Hoc Netw. **81**, 143–159 (2018), http://www.sciencedirect.com/science/article/pii/S1570870518304906

Chapter 3
Developing an IoT Access Control Architecture

Abstract This chapter presents the development of an access control architecture for resource constrained IoT devices. We discuss how to manage policies in a manner that is both scalable and flexible for such IoT systems. We introduce a policy-based approach that provides fine-grained access for authorized users to services while protecting valuable resources from unauthorized access. We use a hybrid approach by employing attributes, roles and capabilities for our authorization design. We apply attributes for role membership assignment and in permission evaluation. Membership of roles grants capabilities. The capabilities which are issued may be parameterized based on further attributes of the user and are then used to access specific services provided by IoT devices. We employ a use-case of a smart healthcare system to show feasibility of the proposed approach in real-world scenario.

3.1 Introduction

The growing number of IoT devices and smart applications in use has created significant potential both for consumers and businesses. But in the wider context of the IoT, it is difficult to manage authorization and authentication both for users and applications. Authorization is the process of granting access rights/privileges to resources and authentication is the act of confirming the identity of an entity. For instance, to unlock a smart door using a mobile phone, the system (and the home-owner) must be reassured that the user is authorized to do so that has preceded by a successful authentication. However, as described in Chap. 1, it is fundamentally challenging for entities in the IoT to always know in advance the identities of other entities with whom they will be interacting.

The challenge is not simply to develop mechanisms and standards to authenticate IoT actors, it is more to authenticate a user or device considering the specific characteristics and the context of such systems. Previously, many computing applications and services were designed within a particular system environment, protected and operated by a dedicated network infrastructure. With the IoT, such systems become open and easily accessible, with users employing their own devices to access system resources e.g. sensors, devices and data [1]. With the increasing number of IoT

S. Pal, *Internet of Things and Access Control*, Smart Sensors, Measurement and Instrumentation 37, https://doi.org/10.1007/978-3-030-64998-2_3

devices (as it is predicted that the Internet will include 50 billion connected devices by 2020 [2]), there will be an impact on a wide range of application sectors as it moves from a network-centred approach to an open one, for example, with remote monitoring of smart sensors, data collection and analysis and seamless device integration surrounding users and their various operations [3].

Consider a practical example of connected healthcare devices in an IoT-enabled smart healthcare system. The use of the IoT has led to a reduction in the strain on medical professionals and improved the patients' well-being while being treated at medical facilities or via remote health monitoring [4]. Specific applications can be seen in many areas from wearable devices to connected beds. For example, the 'NHS test beds' [5] are IoT-enabled smart connected-beds that are used in the UK's National Health Service (NHS) that monitor patients and track data. This is effective in enabling elderly patients to monitor their long-term health conditions while admitted to a hospital. Another IoT-enabled device named 'QardioCore' [6], a wearable ECG monitor, can provide continuous medical-grade data of a person to the health centres that monitor conditions e.g. diabetes and heart troubles. Similarly, 'Zanthion' [7] is an IoT-enabled smart medical device that is used to track a patient's movement. It generates alerts if a patient remains motionless for too long a period of time. However, all these applications generate significant amounts of sensitive and private data.

From tablets, smartphones to thermostats and smart metres, the question lies in building secure authentication and authorization systems that in turn will make the data more secure. Demand for the use of IoT-enabled solutions is only likely to grow, with users relying on this technology for efficient and secure access to personal data.

3.1.1 Problem Description

While such a convergence of digital-physical systems can provide better services, reduced cost of applications and improved user experience, it also leads to numerous challenges in security and privacy [8–10]. Attacks on IoT systems are becoming more sophisticated [11, 12]. This is not limited to simply infecting network traffic with malicious code. IoT-enabled devices can be compromised and controlled remotely by the attackers. For example, a patient's pacemaker can be used to generate a fatal shock or a drug infusion pump (e.g. insulin or antibiotics) can be controlled by an attacker to change the drug dosage [13]. Unfortunately, the characteristics of the IoT, e.g. low-powered devices, small memory capacity and limited processing power, are major issues that impact on the creation of secure IoT systems [14]. This means that it is impractical to enforce heavy-weight security mechanisms in these devices while the scale of the systems argues against fully centralized solutions. From a communication point of view, heterogeneous network environments, wireless communication mediums, high mobility of *things*, dynamic network topology and availability of infrastructure for communication present significant issues [15]. Further, from the access point of view there are several use cases that need to be considered. For instance, a user wants to access an operation from a *thing* for the first time or many times for the same operation. An access could also be made for different operations at the same time or in a certain interval of time. Given the dynamic and the high mobility in such IoT systems, an access control framework should address such issues with proper authorization and authentication.

In a large-scale IoT system, with its open technologies and resource-constrained IoT devices (with limited battery power, memory capacity and computational speed, etc.), managing the resources and users of the system and enforcing appropriate policies are complex and challenging issues [16]. IoT systems can be very dynamic with ever-evolving environments. These systems can also not afford single point of failure risks from over-centralization but conversely require high-levels of security due to the sensitive and mission-critical nature of these systems and the scale of data to be handled. These, and other characteristics, necessitate solutions that are specifically designed for the IoT arena. In particular, identity, access control and privacy have been identified as pressing issues in this context [17]. IoT systems will need policies and mechanisms to support authorization (i.e. determining whether an entity can access a resource) and authentication (i.e. identifying an entity). With the range and scale of users, applications and data in an IoT setting, the scale of the policies required for access control, and the management of those policies, must be considered.

As we discussed in Chap. 2, a number of well-known access control models and mechanisms have been proposed for use in the IoT. In this section, we quickly recap three of the major access control mechanisms that are related to our research, RBAC [18], ABAC [19] and CapBAC [20]. As noted in the previous chapter, each of these mechanisms has its own advantages and disadvantages when applied to the IoT. RBAC provides fine-grained access control over the resources using explicit user-to-role mappings, however, RBAC itself is highly centralized and requires the definition of each user-to-permission relation for each resource that a user is to be allowed to access. This is challenging in a large and complex system like the IoT [21]. ABAC improves policy management by using attributes (e.g. name, age, location, etc.) rather than concrete identity. This is more flexible for the IoT as policies can be written based on the context (e.g. current time or a location). However, ABAC by itself provides no mechanism for controlling the number of policies required, e.g. by grouping together policies with the same attribute requirements. This is an issue in highly scalable systems e.g. the IoT [22]. CapBAC provides flexible access control. Users are provided with capabilities which identify the resources and operations on that resource that a user is allowed to access. This allows fine-grained access control. However, many of the existing (non-IoT) CapBAC mechanisms are centralized when validating access rights of subjects. Distributed CapBAC models e.g. [23, 24] have been proposed for use in the IoT. In Distributed CapBAC models, validation is performed inside the resource-constrained IoT devices (or a local management capacity) without there being any contact with a centralized authority. This allows a distributed approach, taking advantage of edge-intelligence, i.e. in-line with the nature of the IoT. However, these systems do not address the problems of managing the number of capabilities that will be required in a realistic IoT system, let al.one the policy base needed to control their creation and distribution. The policy base that will need to be defined is likely to be large and dynamic. The scale and diverse nature of the IoT makes it difficult to specify, centrally and in advance, a complete set of access control policies.

Now, let us consider the following example situation, which is likely to be common in an IoT-enabled healthcare setting. We use healthcare as an example as it is a representation of a dynamic IoT environment, with a large concentration of devices and highly sensitive data. A number of medical sensors (e.g. to monitor blood pressure, body temperature, etc.) are attached to a patient. The patient's doctors should be given access to the sensors to allow readings to be taken. Defining policies which give access to each doctor for each sensor will be time-consuming and hard to manage. Given the dynamic nature of both the IoT and healthcare situations, the set of sensors is likely to change in unpredictable ways, making managing their access particularly challenging. The problem of policy management, and particularly the number of policies that must be authored, requires addressing. In short, in the context of the IoT the common problem with the aforementioned access control mechanisms is that, in isolation, they are not explicitly designed or suitable for addressing scalability, whether it is in terms of devices, users or policies.

3.1.2 Contributions

The aim of this chapter is to present the design of an access control framework for the IoT and at the same time to reduce the number of policies required for authenticating a legitimate user within the system that may include thousands of users and *things*. In particular, we make the following contributions:

- We design a policy-based, fine-grained and partially decentralized access control architecture based on attributes, roles and capabilities.
- We use attributes to parameterize capabilities for accessing specific services provided by IoT *things*.
- We provide a detailed architecture, system design, formal model, implementation and evaluation of the system prototype.
- We examine the use of light-weight network authentication protocol (e.g. symmetric key based approach) for resource constrained IoT devices.
- We provide a detailed discussion of the employed symmetric key based approach in comparison with the asymmetric key based approach. Note that the architecture could easily work with either approach.

3.2 Background

In Chap. 2, we provided a detailed discussion of access control issues and mechanisms and a comprehensive analysis of access control in the IoT. In this section, we outline a few existing proposals for IoT access control [25–28]. For instance, Sahi et al. [29] discuss the technical requirements for secure and controlled access control in smart IoT environments. Yeah et al. [30] propose a fine-grained health information access control framework for light-weight IoT devices. Gandhi et al. [31] present a detailed survey on intelligent access control for healthcare systems using the IoT. Other contributions have surveyed specific areas and technologies related to IoT and smart healthcare systems [32–36]. Our intention is not to discuss IoT-enabled smart healthcare systems in detail. We simply exploit this use case to demonstrate the usefulness and popularity of IoT-enabled systems, as one of the examples, in real-world scenario. In this section, we quickly recall an architecture for IoT in which our proposed access control framework would perform. Next we provide a brief discussion to the use of RBAC, ABAC and CapBAC for the IoT. Then we outline the current limitations of access control to an IoT system.

3.2.1 IoT-Enabled Smart Systems

The vision of an IoT-enabled smart system is to operate autonomously while seamlessly connecting the various devices, sensors and human users [37]. The use of technology and communications are however varied according to the system's requirements and designer's choice. A typical IoT-enabled smart system consists of wearable sensors, advanced communication techniques and diverse network connections [38]. This can be seen as a multi-layer architecture. Recall, from Chap. 2, the various reference models for the IoT. In this section, we adopt a simple four layer architecture composed of a sensing layer, network layer, service layer and application interface layer. In Fig. 3.1, we depict these four layers. In the sensing layer, sensors are used to automatically detect an environment's condition and collect data from different physical devices. The major sources are RFID tags and intelligent wireless sensors. The network layer connects the various sensors with different networking connections and is capable of transferring information from the sensing layer to the service layer. The service layer provides a cost-effective middleware platform for service aggregation, service composition and service division, etc. In this layer advanced algorithms can be used to process data and analyse them according to the end-user requirements. Finally, the application interface layer helps users to access services using Application Program Interfaces (APIs) or apps.

Refer to our use case example of an IoT-enabled smart healthcare system, sensors (wired or wireless) are attached to patients to monitor their health data and periodically collect them and transmit this information in real-time (sensing layer). The collected data are then stored in a healthcare-facility database via a suitable com-

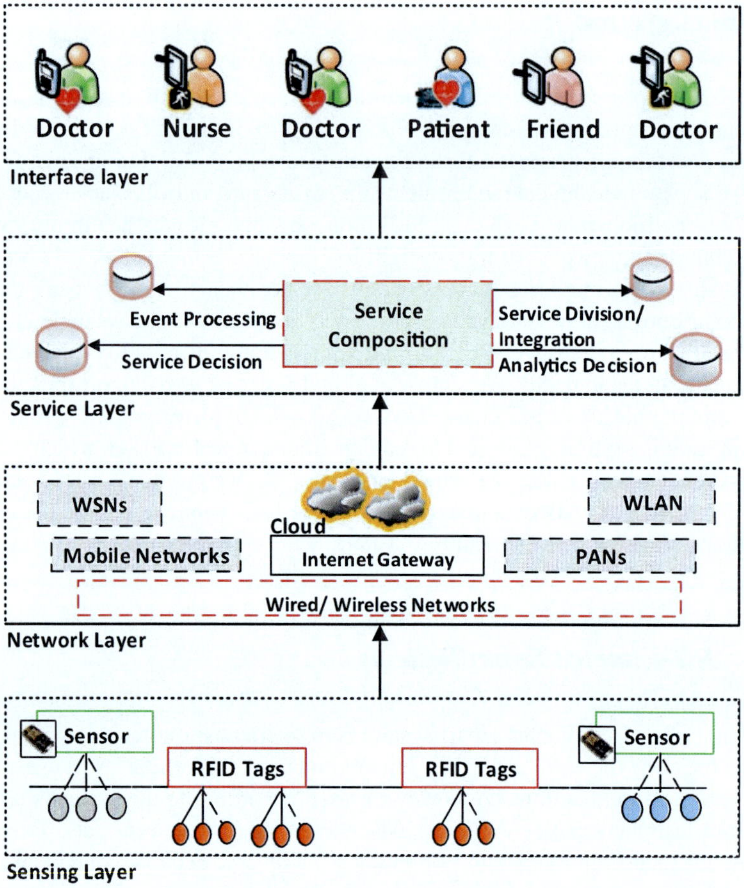

Fig. 3.1 A layer view of an IoT-enabled smart healthcare architecture

munication protocol to a gateway e.g. via Personal Area Networks (PAN) or WSN protocols (network layer). The stored data are analyzed, evaluated and sent to suitable entities for further action (service layer). Finally, authorized staff (e.g. doctors, nurses, etc.) can view the appropriate patient's data (interface layer).

3.2.2 *Current Limitations*

Recall, in Sect. 2.3.3 of Chap. 2, we described commonly used access control mechanisms e.g. RBAC, ABAC and CapBAC for the IoT in detail. In this section, our aim is to examine the limitations of existing approaches from two perspectives, first, flexible and fine-grained policy management and second, securing communication and authentication, given the resource constrained nature of the IoT *things*.

3.2.2.1 Flexible and Fine-Grained Policy Management

Access control in the IoT requires the inclusion of proper policy enforcement mechanisms, which must define how the system should interact with other systems and entities. Recall, policies can be seen as the operating rules that regulate data management, data consistency and security in general. This is achieved by the underlying security policies determined by the policy decision component of the system. This can vary from system to system. Policy management in the IoT is crucial due to the dynamic nature of the IoT and federation of cross domain architecture. For instance, enforcing policies across domain boundaries or over multiple domains. This further enhances the specific requirements of the domains and their policy settings. For example, in an IoT-enabled smart healthcare setting that is spread over multiple domains, cooperation and communication between different actors, clinical establishments and hospital management are crucial when communicating with one another. Policy management in the IoT has not received significant attention in the existing literature.

Further, each of the aforementioned mechanisms (i.e. RBAC, ABAC and CapBAC), in isolation, has its drawbacks when enforcing policies for a large-scale dynamic system like the IoT [39, 40]. RBAC provides effective policy management but it is dependent on a highly centralized system. It also typically requires explicit user-assignment to roles, and supports only predefined and static policies. This is inflexible in a highly dynamic context like the IoT. ABAC improves flexibility in policy management as explicit user identities are not specified in the policies, rather users are identified based on attributes they possess. The use of attributes in ABAC assists in the enforcement of fine-grained access control policies in real-time [41]. However, there are questions around how many policies are required and where they are evaluated. If one set of attributes is to give access to multiple resources, and this evolves over time, either multiple policies, or significant policy re-writes, will be required. The resource-constrained and highly distributed nature of the IoT raises questions of where the attribute policy base is stored, and its policies checked and how many copies of those mechanisms may be required if the users' attributes need to be consulted on each access to a resource. CapBAC simplifies the distribution of permissions and is decentralized by nature. However, previous proposals for the use of CapBAC in the IoT have ignored the question of policy management at a fine-grained level. Further, the issues of capability propagation and revocation are two major challenges in large-scale IoT systems.

3.2.2.2 Securing Communication and Authentication

Several approaches have discussed communication and authentication issues for IoT-enabled smart systems. For instance, approaches e.g. [42, 43] discuss user's anonymity and the use of two-factor user authentication to prevent security vulnerabilities e.g. replay attack and password data disclosure. However, both of these contributions are highly centralized and use PKI-based system, which are not ideal bases for IoT systems. The characteristics of IoT devices impose challenges where heavy-weight security mechanisms cannot be employed directly [44].

From the protocol point of view, there are multiple options that could potentially be adopted in resource-constrained IoT networks. On the one hand, new communication protocols can be implemented for the IoT on top of the IPv6 infrastructure. On the other hand, existing light-weight protocols could be used to provide interoperability with the existing infrastructure. The latter provides flexibility when integrated into the existing architectures [45]. We argue that, to achieve light-weight authentication by verifying the identities of principles for the IoT, protocols e.g. Kerberos [46], could be an alternative. It provides a light-weight protocol (using symmetric-key cryptography) that achieves both authentication as well as authorization (without using public key cryptography). Kerberos uses access tokens (commonly known as *tickets*) to authenticate clients and servers and grant access to services [47]. It resolves authentication, digital signature and encryption in a single sweep. Since the IoT *things* are ephemeral, a Kerberos-like communication protocol could be suitable for various related applications. It could also be integrated with a light-weight application level protocol e.g. CoAP [48]. However, extensive efforts are lacking on how to employ light-weight protocols for achieving authentication in large-scale IoT systems.

3.3 An Example Scenario

In this section, we discuss a practical use-case scenario. In a healthcare context, IoT devices can be used to monitor, collect, view and analyse patient medical data in real-time and control and monitor access to the data, patient services (e.g. drug administration) and the physical building. The devices may communicate with each other. As examples, some devices would collect patient data and make them available to appropriate parties. Other devices, e.g. those carried or worn by patients, staff and visitors, would allow authorized access to data and to building facilities. In Fig. 3.2 we depict a generalized IoT-enabled healthcare system where several actors are involved. The following actors appear in our scenario:

- Patient: A person who seeks a medical treatment.
- Doctor: A medical practitioner who provides general medical treatment.
- Specialist: Specialized medical practitioners, e.g. cardiologists, who have training in, and provide treatment for particular medical conditions.
- Nurse: Clinical personnel who provide care and treatment of patients.
- Staff: Technical staff of the healthcare institutions, e.g. IT support.
- Family: Relatives, friends and other visitors of the patient.

Actors within the system may require repeated access to a resource. For example, nurses and doctors attending a patient are likely to require continued access to the

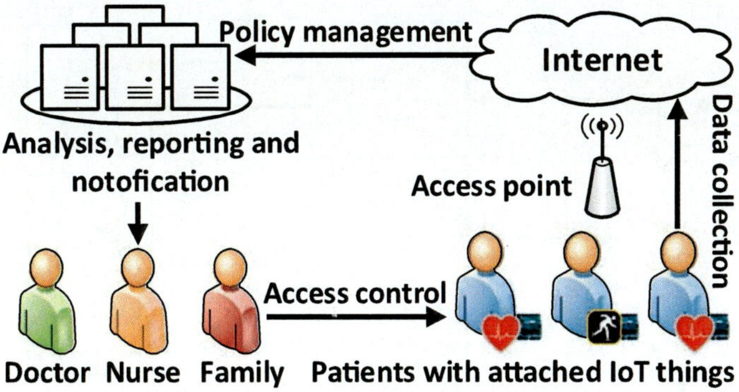

Fig. 3.2 Different actors and the flow of information in an IoT-enabled smart healthcare system

sensors attached to the patients. All actors will have parts of the building and its infrastructure (e.g. the doors, lighting and air-conditioning of rooms) to which they have authorized access and will repeatedly access those parts through the devices that control such access. It would be preferable if the access to such devices could be authorized locally, without recourse to repeated communication with a central access control system. Note also that much of this access will be to classes of users (e.g. nurses, doctors, patients) and not adjudicated on the basis of individual identity. Authoring and managing policies on an individual basis would be unwieldy. Even on an attribute basis, separately specifying access to different resources based on the same attributes (e.g. doors, lights and the sensors of patients in a particular ward) would lead to an unwanted number of policy expressions.

Staff at the facility may be assigned rotating duties, for example, nurses may cycle through different wards. Rather than explicitly modifying the assignment of permissions to users (which may, in policy terms, be expressed as role membership), the change could be reflected by an update in the recorded attributes of the actor. The attributes then grant appropriate access, by being used to judge role membership. If the requirements for role membership (and hence the actual membership) change then this can be expressed as a single change (in the attribute assignment) rather than multiple changes (in the user to role assignment), simplifying policy management.

Specialized staff will require access to specialized equipment as well as devices attached to certain patients. For example, a cardiologist should be given access to the output of the heart monitors of the patients under their care, and no others. Simply giving all cardiologists membership of an appropriate role (e.g. *cardiologist*) which grants access to devices of type *heart_sensor* grants unnecessary access, as members of the role will be able to access all such devices. Creating a role for each patient (e.g. *cardiologist_of_Bob*) and mapping the cardiologists as needed would needlessly multiply roles (policies) and increase the difficulty of policy management. By recognizing the doctor-patient assignment as an attribute it is possible to ensure that the access provided (via a capability) is only to the appropriate patient. This could still be achieved via a single cardiologist role, if the capabilities obtained

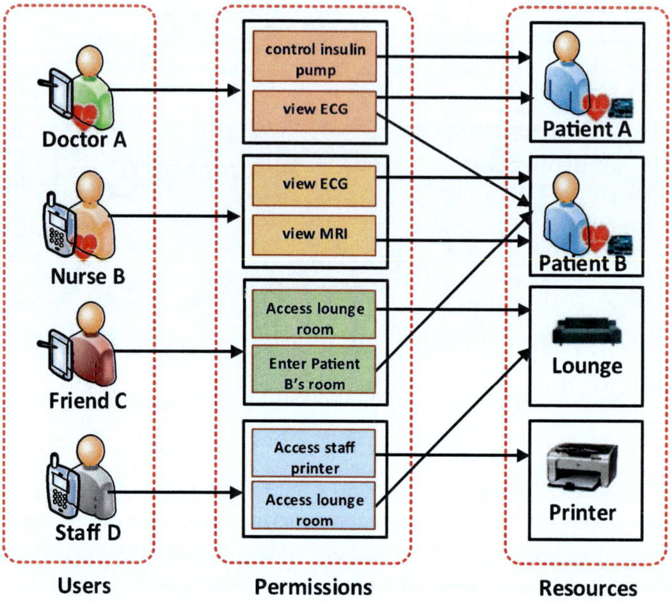

Fig. 3.3 The vision of a fine-grained and controlled access control scenario for different actors in a healthcare system. Where a user with an appropriate permission can only view or control a subset of resources. For instance, Staff D can access both the lounge and printer, but Friend C is only allowed to access the lounge room while visiting Patient B

through membership of the role are tailored (parameterized) according to the doctor-patient relationship (as defined by the relevant attributes). We can allow access for the cardiologist to only their patient's sensor with a small number of policies. This approach also allows for rapid allocation of staff to patients and for changes in the sensors attached to the patient, as only the attributes of the actors require updating and not the access control policy expressions themselves.

During their time in the healthcare institution patients are likely to change ward. For the most part this is similar to the case above that dealt with rotation of staff. However, many patients will have friends and family who visit them. The hospital would allow these registered visitors access to the patient's room, and possibly other appropriate parts of the facility. The required access changes as the patient's status within the institution changes. If the visitors have an attribute with the meaning 'visitor of patient Bob' this can be used both to grant them membership of an appropriate role (e.g. *visitor*) and to tailor the capabilities with which they are provided based on the status of patient Bob. Note that the access then granted to visitors varies as attributes settings of the patients vary (assuming appropriate policy settings in role *visitor*) without any changes to either the visitors' attributes or any policy settings. This again simplifies policy expression and management. In Fig. 3.3, we show an example of the different users and their various access permissions to various resources.

The above use cases, and similar considerations, lead to our proposed design. Access is provided by granting capabilities. Capabilities can be checked locally, by the smart devices or their local management, without recourse to the central system, and can be used repeatedly, again helping to limit reliance on a central system. Attributes provide a powerful method of specifying access policies in a flexible and fine-grained manner. Using attributes to define role membership, with role membership specifying which capabilities are available, reduces the number of policies that have to be created by allowing a single attribute expression to provide access to multiple resources. Finally, by using user attributes to tailor the capabilities issued different members of the same role can be provided with access to different resources, again reducing the number of policies required and simplifying policy management.

3.4 Proposed Access Control Architecture

In this section, we discuss our proposed access control architecture. First, we outline a list of assumptions and then illustrate architectural components in detail. We present a formal specification of our model followed by a discussion on the core modules of the system. We also present the capability structure and the process for its instantiation.

3.4.1 Assumptions

To develop our access control architecture, we make the following assumptions specific to our design.

- Users have a smart device.
- Smart devices broadcast services to the other users and devices.
- Devices are not being hacked or compromised by attackers.
- Users (e.g. employees in a certain organization) do not disclose or lose their unique ids generated by the administration.
- Users supply appropriate and correct policies based on the principles of the system administration.
- Attributes are properly assigned to the *things*.
- User and device attributes are not miss-assigned.

3.4.2 System Functionality

Before we provide a detailed discussion of the proposed access control architecture, let us consider how the system would function. In Chap. 1, we have noted that in an

IoT system, communications between *things* can involve various device types (e.g. smart phones and Internet-aware devices), routing protocols (e.g. RPL, etc.) and interaction patterns. We argue that the information interchange between IoT *things* can be initiated in different ways. In our case, we assume that the *things* broadcast services, within a close proximity. In other words, the service that the *things* provide to users and other devices are physically present in a short distance. The *Google Beacons* [49] platform is an example of such technology. It broadcasts location and service information by typically in the form of a Bluetooth Low Energy (BLE) beacon. Note, nearby notifications help users discover the services and applications that surrounded them, by surfacing location-specific notifications for apps and websites. In such cases, sometimes a receiving app is required. In this chapter, we focus on abstracting the usability, service discovery and interactions patterns between the users and the *things*. Note, we reference Google Beacons to exemplify the potential interaction patterns between the users and the *things* in various architectural approaches.

Let us assume that the *things* are registered in the system via a central management system (e.g. a central registration server). They (i.e the *things*) broadcast services with the list of possible APIs for users within a communication range. The user receives the list of APIs (on their mobile devices) and wants to perform one of the operations from the list. To perform the desire operation, the user requires a valid capability. To obtain the capability, the user communicates with a central management system and checks whether it is allowed to access that particular service. The user also informs the central management system of the identity of the *thing* and the operation the he/she wishes to perform on the *thing*. After receiving the request from the users, the central management system determines the user's role-membership based on attributes. It also checks if the *thing's* registration is valid. If the *thing's* registration (including the requested operation) and required attributes are valid, then the central management system proceeds towards a capability generation stage based on the policies. Otherwise (i.e. if *thing's* registration or attributes are not valid), it terminates the service request. The capability is issued from the capability templates for the requested operation with the user's identity embedded. The issued capability is now forwarded to the user. The user is now able to present the capability to the *thing*. If the capability is valid (including testing any conditions included with the capability), then the *thing* grants the requested operation. Otherwise (e.g. the capability is not valid), the service request will be terminated.

Note, the notion of attribute is central to ABAC systems e.g. ours. We mainly distinguish two types of attributes depending upon when they are required to have well defined values.

- *Static Attributes*: In ABAC, the evaluation of a policy ultimately depends upon attributes that are known at the time of processing the access request and in particular, during the policy evaluation. We refer to these as *static attributes* [50]. Commonly, the values of a static attribute are set once within a given context and such values either do not change afterwards or the values have a very long lifetime.

Examples of static attributes include unique identifiers, RFID card number, name, age, phone number, resource id, date and time of a request, etc.
- *Dynamic Attributes*: If an attribute cannot be assigned to a fixed value at the evaluation of the policy then we refer to it as a *dynamic attribute* [51]. These attributes play an important role in CapBAC. A *dynamic capability* is a capability that depends upon one or more dynamic attributes.[1] Access time is an obvious example of a dynamic attribute. Another example is the access location e.g. *a capability that unlocks a door of a patient's room must be used within a close proximity to that door*. Such a condition can be built using a *proximity-attribute* which can be implemented using a proximity challenge.[2] Dynamic conditions cannot be evaluated at the time of authorization or evaluation of the policies that govern issuing of capabilities.

3.4.3 Granting Different Level of Access

In our use case scenario (discussed in Sect. 3.3), we argue that we want to provide different actors with different levels of access to the patient's medical sensors and the data they provide while protecting the patient's privacy. For simplicity, we consider that the *things* for different patients may have different associated access control policies and that these policies are stored inside a centralized system. However, enforcement is handled locally at the edge devices. This may either be within the *things* themselves or a local management device if the *things* have insufficient functionality for the task. This both takes advantage of the edge-intelligence of such systems and avoids performance bottlenecks in the central system.

Consider, a real-life hospital environment, where there may be hundreds of doctors, nurses and other staff and possibly thousands of patients with each patient having multiple sensors attached. No one medical professional should have access to all sensors on all patients. Conceivably, no medical professional may be able to access all the sensors on a single patient and even if they could the levels of access may vary from medical professional to medical professional. Consider two examples of the complexities derived from the aforementioned use case:

- Medical specialists, e.g. cardiologists, should only be able to access the relevant information for patients under their care. If Doctor A is treating patient Alice and Doctor B is treating patient Bob then A should only be able to access the readout from Alice's heart sensor and B should only be able to access the readout from Bob's heart sensor.
- Nurses may be assigned to a particular ward. Each patient in the ward may have a standard set of sensors attached, in addition to ones that may be particular to their

[1] Capability conditions are the standard implementation of dynamic capabilities.

[2] Intuitively, the resource checks how close the requesting subject is located (i.e. to determine its proximity) by sending a challenge to the subject. The solution to this challenge may require short range communications or even a physical interaction.

condition. Nurses should be able to access the standard sensors for all patients in their ward, but not for patients in other wards.

3.4.3.1 Policy Management

Within this general system description above there remains a question as to how to formulate and distribute capabilities while maintaining the minimum number of access control policies. Consider how policies may be written, and permissions granted (i.e. capabilities distributed). A number of alternatives exist:

1. A simplistic solution would be to create a role 'cardiologist' which provides its members with a capability that grants access to all sensors of type 'heart monitor'. This would allow every cardiologist to access every patient's heart monitor, violating patient privacy.
2. Each capability could have associated with it a test, evaluated on access, that ensures the patient is under the care of the specialist presenting the capability. This would increase the processing and bandwidth requirements on the *things*. The relevant credential, proving the relationship between specialist and patient, would have to be provided to the *things* and any signature on it checked. Signature checking is the most time-consuming activity involved so any such extra requirements on the *things* should be avoided, especially as the check would need to occur on every access.
3. Patient-specific roles, e.g. 'cardiologist of Alice' and 'cardiologist of Bob', could be created, which would only confer access to the sensor(s) of the particular patient. This would fulfil the requirements for specialists to only be granted access to the sensors of their patients. However, it would be difficult and time-consuming to manage and produce a large number of similar policies. It may also be difficult to assemble this information a priori (given both the large number and dynamic nature of doctor-patient combinations).

None of these alternatives provides us with the required level of flexibility and fine-grained access without creating a needlessly high number of policies to be authored and maintained.

A preferable alternative is to take the first option above, but provide additional information, e.g. a credential proving the relationship between specialist and patient, to the capability issuing system. The capability generated and returned to the user would then only grant access to the nominated patient. In effect, the capabilities are parameterized by the additional information provided, in a manner analogous to the role parameterization of [52]. While the signature on the attribute credentials may still have to be checked, this is superior to option two as the signature is only checked once (on capability issuance, not on every use) and by the central policy management system, which will have superior processing power compared to the individual *things*. The *things* would only have to check the signature on the capability, not on both the capability and the attribute credential.

The solution also fulfils the requirements of the second example given, although here the information provided for parameterization would be a credential affirming assignment of a nurse to a particular ward. In both cases the *things* would need to be registered with the central system, along with such attributes as the patient they are assigned to and that patient's current ward.

Note that in effect the capability issuing system stores *capability templates* as defined by the relevant policies. Most CapBAC systems for IoT access control would effectively store capability templates at least in a simplistic manner. For example, the capabilities of [20] include expiry time and identity of the user to whom the capability is issued. These would be place-holder values in capability template stored by the policy manager and filled in when the actual capability is instantiated for distribution. We extend this idea to the identity of the resource to which the capability allows access. For example, the template might note that the capability allows to devices of class *'heart_sensor'* but that the issued capability can only give access to the heart sensor of a patient under the care of the requesting actor. Proof of this relationship between specialist and patient would be an attribute credential demonstrating its existence. The heart sensors would be registered by the system as being assigned to the patient. From this information and the capability template an appropriate capability can be generated and issued.

The effect of this approach is analogous to that of Schwartmann [53], who was also concerned with providing fine-grained access within a healthcare system. However, Schwartmann's approach involved predefining activation contents for *each* patient on a per-role basis. This does not scale well as the number of patients increases, as the policy expression for each patient must be handled independently.

As we demonstrate in Sect. 3.4.6, by relying on attributes attesting to the relationship we can abstract such policy settings into a minimal number of actual policies. Further differences between our system and that of [53] include the reliance of the latter on explicit user assignment to roles and its centralized approach to permission checking (i.e. the central system is consulted on every access) rather than employing capabilities.

3.4.3.2 Capability Management

Note that in effect the capability issuing system stores *capability templates* as defined by the relevant policies. On issuance, the templates are filled in with relevant information, e.g. validity time, and the appropriate parameterization information. Details on the different fields of a capability and their instantiation are discussed in Sect. 3.4.7. Here we only outline a general discussion on how a capability is issued.

Figure 3.4 illustrates a simple outline of a capability issuing process using a capability template. When a request for a capability reaches to the capability issuer (step 1), it checks the corresponding access policies from the policy database (step 2) and if satisfied (step 3) it uses appropriate capability templates (step 4) to instantiate a capability from the corresponding capability template storage (step 5). Finally, it sends the issued capability to the user (step 6). If the corresponding policies do not

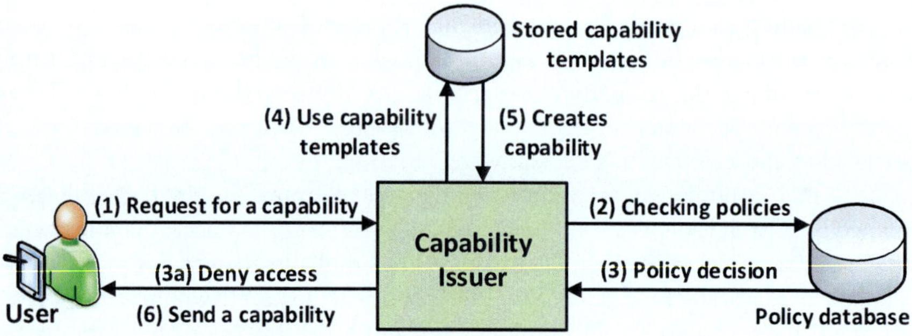

Fig. 3.4 Issuing of a capability from a capability template

match, then the request terminates at the first instance and a response sends back to the user (step 3a).

3.4.4 *Overview of the Architectural Components*

In Fig. 3.5 we depict our proposed access control architecture. The proposed architecture consists of the following main components: User Devices (UD), Things (TH), Central Management System (CMS) and Things Registration Repository (TRR).

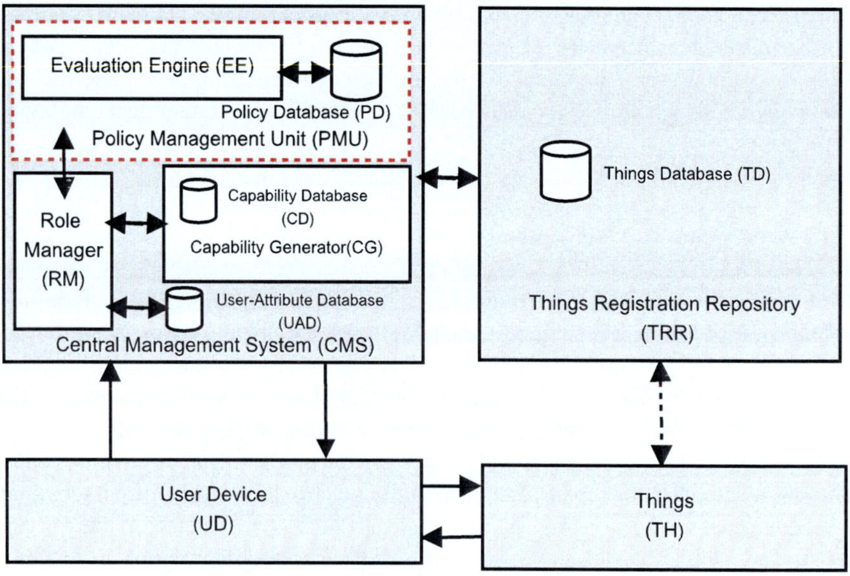

Fig. 3.5 The proposed access control architecture

The CMS consists of the Role Manager (RM), Capability Generator (CG), User-Attribute Database (UAD) and Policy Management Unit (PMU). The PMU includes an Evaluation Engine (EE) and Policy Database (PD). The CG contains the Capability Database (CD). The TRR holds the Things Database (TD).

A UD is a smart mobile device (e.g. smart phones, tablets, PDAs, etc.) belonging to a system actor (user). We assume that users are authenticated (using appropriate attributes) to their corresponding UDs, i.e. the UD knows who the user using it is. The UD also stores capabilities issued to its user. The generation of attributes is out of the scope of this research and we assume that the attributes are generated by a trusted authority. TH includes both the smart IoT *things* (e.g. a heart monitor sensor) and local security management devices. For example, smart sensors attached to the patient's body. We assume that THs, or their local controllers, have the ability to store the long-term key associated with the TH and to check supplied capabilities. Importantly, in our architecture, a TH is unaware of the user's roles and attributes. This improves the user's privacy and limits the functionality required of edge security devices.

In our design, the CMS is a centralized component. The CMS acts as a central server that provides local area networking services to multiple users at a time. It could be one or a combination of multiple high-speed computers for storing and processing data files which can be shared by different users. The major functions of the CMS include role assignment, role authorization, permission authorization and issuance of capabilities and provision of session's keys for communication between the UDs and THs. It is infeasible to widely distribute copies of the CMS within the resource-constrained IoT devices, therefore, we situate the CMS in the central server. The RM holds the role hierarchy and coordinates the activities of the other components of the CMS. The PMU is used to verify these expressions against the actual attributes of a user, retrieved from the UAD, to determine if the user should be allowed the access specified by the role. The EE evaluates a user request by locating the attribute rules that must be satisfied for role membership. The EE holds attribute rules which grant role membership and define capability parameterization. The policies are stored based on a policy language, e.g. XACML. Recall, XACML is an XML-based general purpose access control policy decision language for managing access to resources.

Access to TH's is represented as permissions associated with appropriate roles. These permissions are capability templates, which can be instantiated with appropriate information to form actual capabilities for distribution to users. Role membership is specified as an attribute expression, not by explicit user assignment. The CD stores the capability templates, whose instantiation can be governed by policies managed by the PMU, and the mapping of capability templates to roles. Capabilities are generated from the capability templates. This will include, for example, inserting the user identity into the capability and the issuance and expiry times. More details of capability generation are given in Sect. 3.4.7. The CD also stores the revoked capability lists. Lastly, the TRR manages the valid registrations (e.g. identities, etc.) and attributes of the THs (stored in the TD). It is dynamically updated when a new TH joins or when an existing TH leaves the network.

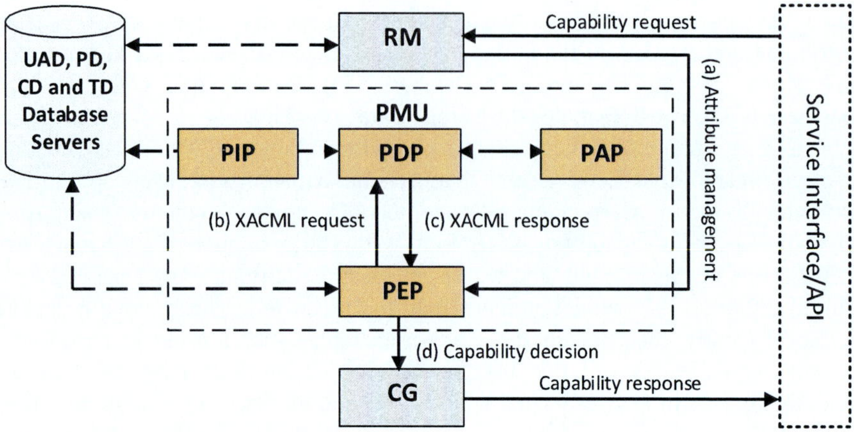

Fig. 3.6 The core modules of the system design

3.4.5 Core Modules of the System

In this section, we discuss the core modules and functioning of the system (i.e. the core modules of the CMS). In Fig. 3.6, we depict the core modules of the system. Importantly, Fig. 3.6 expands upon Fig. 3.5, illustrating the modules of the CMS.
Role Manager (RM): This module is responsible for mapping role membership based on the attributes. Note that users are not directly assigned to roles, their membership is judged on the attributes they possess when an access request is made. The roles are typically ordered hierarchically. This is particularly important when writing a policy that should be inherited by other roles (e.g. a doctor can access everything a nurse is allowed to access).
Database Servers: This includes the UAD, PD, CD and the TD. These databases can be managed on possibly different servers. They can be in the form of conventional relational databases (for policies and THs) as well as specialized storage e.g. Active Directory for user management.

The PD stores policies in the form of serialized XACML policies and the associated metadata. XACML is a widely used modern standard for writing fine-grained access control policies. The PEP (Policy Enforcement Point) and PDP (Policy Decision Point) engines described below communicate through the publication of XACML requests and responses. The PD also contains auxiliary resource tables that store templates required for the generation of these XACML messages. The UAD stores user attributes e.g. name, ID, hashed password, phone number, etc. The PEP engine uses this database to authenticate and verify the user's attributes. The TD stores attributes of currently enabled and deployed THs. As noted above, the CD stores the capability templates. A template can be referred to as the default pre-build module that is used in the generation of an actual capability (i.e. instantiation of capability) and is composed of specific fields as needed.

Policy Management Unit (PMU): The PMU has the standard XACML architecture to provide a cross-platform solution while satisfying interoperability. The following components are part of the EE and responsible for creating, evaluating and responding to requests based on the policies [54].

(1) Policy Enforcement Point (PEP): This engine translates queries into syntactically correct XACML requests. This is achieved by combining the stored user attributes (from the UAD) and service request details with the details of the addressed resources, contextual data, and in particular timestamp (step a of Fig. 3.6). The contextual details are accessed using the PIP (Policy Information Point) module described below. The XACML request is then forwarded to the PDP engine for evaluation (step b of Fig. 3.6). The response from the evaluation engine is in the form of an XACML response (step c of Fig. 3.6). This response is translated by the PEP into a format that is understood by the CG (step d of Fig. 3.6).

(2) Policy Decision Point (PDP): This engine receives the XACML request from the PEP and evaluates it against the access control policies. The active policies are loaded from the PD. The PDP consults the PD to determine which role(s) gives access to the requested resource. Stored with the role is the attribute policy which governs membership of that role. That policy is also retrieved at this point. The evaluation process then evaluates the policy against the attributes in the XACML request. When multiple rules apply to the request then a combining algorithm is required to resolve any conflict arising. In our example implementation system, we use 'Deny Override' to provide a very restrictive access control but this is highly configurable at the policy level. The result is then sent back to the PEP for enforcement in the form of an XACML response.

(3) Policy Information Point (PIP): The PIP connects the PEP (and possibly the PDP) to the underlying sources of attributes (i.e. the attribute values). This functionality can be implemented implicitly within the PEP engine which communicates with the UAD and the TRR.

(4) Policy Administration Point (PAP): Note, most ABAC implementations also use a PAP to create, manage and edit the authorization policy or policy sets. This can be an external application designed to facilitate and secure the management of policies. The mechanics of policy administration are out of the scope for this research.

Note, the PDP of an XACML system usually responds to requests with *permit, deny, indeterminate* or *not applicable*. Not applicable means that the request attributes do not satisfy the rule target and indeterminate means the condition evaluation failed at an intermediate step. These are categorical responses and do not convey much information regarding how the capability shall be utilized. Moreover, information about the usage of the capability must be encoded within the policy files themselves so that they can be maintained by the policy administrators. Interestingly, XACML policies allow the encoding of post-decision conditions using *obligations* and it is up to the PEP of an XACML system to interpret and enforce such a side condition. In our case, the PEP enforces such conditions by adding them into the capability conditions.

Capability Generator (CG): This module is responsible for generating capabilities for positive XACML responses (negative responses will be reported as an authorization failure to the requesting user). The CG stores the capability templates inside the CD. Recall, an actual capability is issued from an appropriate capability template. The CD also stores a list of revoked capabilities. However, the revocation of a capability is outside the contribution of this chapter.

3.4.6 A Formal Specification of the Model

Our proposed model has the following components: $R, A, Capt, Cap, U, T, O, Cla$ and E ($roles$, $attributes$, $capability\ templates$, $capabilities$, $users$, $things$, $operations$, $classes$ and $environment$ respectively). We represent a Cla as an extensible programme that creates objects for designing and building applications. We define the following mappings:

- $RCapt : R \times Capt$, a many-to-many role to capability template assignment relation.
- $ClaO : Cla \times O$, a many-to-many class to operation assignment relation.
- $ClaT : Cla \times T$, a one-to-many class to $things$ assignment relation.
- $CaptT : Capt \times T$, a many-to-many capability template to $things$ assignment relation.
- $CaptO : Capt \times O$, a many-to-many capability template to operation assignment relation.
- $CaptCap : Capt \times Cap$, a one-to-many capability template to capability assignment relation.
- $UCap : U \times Cap$, a one-to-many user to capability assignment relation.

Note that equivalents to $CaptT$ and $CaptO$, $CapT$ and $CapO$, exist, which map capabilities to *things* and operations. $CapO$ inherits directly from $CaptO$, with capabilities mapping to the operations defined by $CaptO$ for the capability template from which they were derived. The *things* that a capability maps to via $CapT$ is a subset of the corresponding mapping in $CaptT$, as defined by the parameterization rules and supplied attributes.

- $UA_k (1 \leq k \leq K)$, $TA_m (1 \leq m \leq M)$ and $EA_n (1 \leq n \leq N)$ are the pre-defined attributes for users, $things$ and environments, respectively. Where K is the number of user attributes, M is the number of $things$ attributes and N is the number of environment attributes. We follow the approach of [19].
- The attribute assignment relations ($ATTR$) for user u, $things$ t and environment e are $ATTR(u)$, $ATTR(t)$ and $ATTR(e)$ respectively. Where,
  $ATTR(u) \subseteq UA_1 \times UA_2 \times \cdots \times UA_K$
  $ATTR(t) \subseteq TA_1 \times TA_2 \times \cdots \times TA_M$
  $ATTR(e) \subseteq EA_1 \times EA_2 \times \cdots \times EA_N$
- We use the following four attribute-based $Policy\ Rules$ for our model.
  (1) $Role\ Membership\ Rule$: A boolean function of the user and environment attributes $f(ATTR(u), ATTR(e))$. This exists for each role for $role - to - user$

mapping, and specifies what attributes a user (u) must possess to become a member of the role in a specific environment (e). This can be denoted as follows:
$Policy\ Rule : Role\_Membership\ (u, e) \leftarrow$
$f(ATTR(u), ATTR(e))$

(2) *Capability Parameterization Rule*: A rule which specifies which *things* can be accessed using a capability generated from a capability template ($Capt$) given the provided user attributes. This can be denoted as follows:
$Policy\ Rule : Capability\_Parametization\ (u, Capt) \leftarrow$
$f(ATTR(u), Capt)$

(3) *Condition Rule*: This is a boolean function of the *things* and environment attributes $f(ATTR(t), ATTR(e))$. It decides whether a user (u) can access a *thing* (t) in a specific environment (e). This can be denoted as follows:
$Policy\ Rule : Condition\ (t, e) \leftarrow$
$f(ATTR(t), ATTR(e))$

(4) *Delegation Rule*: A boolean function of the user and environment attributes $f(ATTR(u), ATTR(e))$. This attribute rule specifies what attributes a user (u) must possess if that user is to be eligible to employ a delegated capability. This can be denoted as follows:
$Policy\ Rule : Delegation\ (u, e) \leftarrow$
$f(ATTR(u), ATTR(e))$

Now we see how these four policy rules would work. Let us consider an application in the IoT where service providers make available certain products to groups of users without identifying them individually. For instance, a shop may make specials discounts to particular customer based on their attributes e.g. age, suburb and interests. In such case, the *Role Membership Rule* could easily be employed without the need for unique identification of each user visiting the store. For example, age = 35, suburb = epping and interest = video games. In Chap. 5, we provide a more detailed discussion on the use of attributes to identify an entity rather than depending upon their unique concrete identification.

The *Capability Parameterization Rule* defines the creation of an actual capability from a capability template. Typically, a *Capability Parameterization Rule* specifies the required amount of information that must be contained to form an actual capability (discussed in Sect. 3.4.8). This information could be the details about a user, capability issue time, a *Condition Rule*, etc.

The *Condition Rule* denotes the specific conditions that must be followed at the time of an access to a resource. Conditions are an integral and crucial part of a capability. Conditions expressed by the initial creator of a capability may be defined as applying to all users of that capability, i.e. when it is used by the entity to which it is initially issued and when it is used by any entity to which the capability is delegated. Conditions may also be defined which apply only to uses of capabilities delegated from the initial capability. Capability conditions can be expressed as log-

```
<Policy RuleCombiningAlgId="..:first-applicable" Version="3.0">
  <Rule Effect="Permit" RuleId="..:1">
    <Target>
      <AnyOf>
        <Match MatchId="..:string-equal">
          <AttributeDesignator AttributeId="..:act"/>
          <AttributeValue DataType="..#integer">unlock</AttributeValue>
        </Match>
        <Match MatchId="..:string-equal">
          <AttributeDesignator AttributeId="..:act"/>
          <AttributeValue DataType="..#integer">lock</AttributeValue>
        </Match>
      </AnyOf>
    </Target>
    <Condition>
      <Apply FunctionId="..:string-equal">
        <AttributeDesignator AttributeId="..:trg"/>
        <AttributeValue DataType="..#integer">D</AttributeValue>
      </Apply>
    </Condition>
    <ObligationExpressions FulfillOn="Permit" ObligationId="time">
      <AttributeAssignmentExpression AttributeId="..:until">
          <AttributeValue DataType="..#string">
            [[TIME<01-06-2019 00:00:00]]
          </AttributeValue>
      </AttributeAssignmentExpression>
    </ObligationExpressions>
  </Rule>
  <Rule Effect="Deny" RuleId="..:2"/>
</Policy>
```

Fig. 3.7 A simple XACML policy document with access conditions. Note, the XACML syntax has been simplified to improve readability but this is a fully functional policy document

ical predicates over attribute variables and values. For instance, conditions within a capability that allows user `A` to operate door `D` (for certain action lock and unlock) before 01-06-2019 00:00:00 h, can be summarized as follows:

```
[src:A,trg:D,act:unlock,act:lock,time:01-06-2019
                          00:00:00]
```

Recall, in Chap. 2, we discussed the basics of XACML and noted that an XACML rule contains four major parts, namely, *effect*, *target*, *condition* and *obligation*. In Fig. 3.7, we provide an XACML based policy document to illustrate how flexible authorization and delegation can be achieved using the *Condition Rule*. According to the policy, when a request arrives at the PDP, the target of each rule is evaluated sequentially and the first one that is applicable to the request is fully evaluated. Note, in Fig. 3.7, Rule 1 applies for locking and unlocking actions. Any other request will be denied as per Rule 2. Rule 1 contains a condition which specifies this rule is imposed on door `D`. The target and condition blocks must involve static attributes only. The policy described in Fig. 3.7 has one obligations which shall be fulfilled on permit. This side condition implements a validity date before 01-06-2019 00:00:00 h.

Note that when users want to delegate some access rights to others, they need to generate a new capability from a capability providing that right based on the *Delegation Rule* discussed above. Therefore, the delegated capability must adhere to the *Delegation Rule* specified in the policy expression. To achieve this in the delegation, first the user uses appropriate *Capability Parameterization Rule* to generate the initial capability then delegation is performed. Note that delegated capability may contain the conditions specified by the initial creator of the capability, including any conditions that only apply to the delegated capability. This can easily be achieved by employing conditions as illustrated in Fig. 3.7. In other words, a *Delegation Rule* can be seen as an additional condition imposed on top of the *Condition Rule*. We provide a detailed discussion on access right delegation and *Delegation Rule* in Chap. 6.

3.4.7 Capability Structure

Recall, an actual capability is issued from a capability template. We will show a capability instantiation process in the next section (cf. Sect. 3.4.8). In this section, we illustrate the following capability structure that we have used in our model.

Capability Structure: $< Cap_{id}, U_{id}, Iss_{id}, Iss_{time}, Exp_{time}, t, o, Sig, CoR >$

where,

- Cap_{id} : Capability ID is the unique identity of each capability.
- U_{id} : User ID is the unique identity of the specific user to which the capability has been granted.
- Iss_{id} : Issuer ID is the unique identity of the entity issuing the capability.
- Iss_{time} : The time at which the capability was issued to the user.
- Exp_{time} : The time at which the issued capability will expire.
- t : This identifies either a class ID ($cl \in Cla$) or a set of related THs ID, where all the THs are instances of the same class. Note, $t \subseteq T$ and,

$$t = \begin{cases} cl_{id} \mid cl_{id} \in Cla \\ \{t_{id_1}, t_{id_2}, t_{id_3}, \ldots, t_{id_n}\} \mid \exists\, cl_{id} \in Cla \rightarrow \\ \forall t_{id_i} \in \{t_{id_1}, t_{id_2}, t_{id_3}, \ldots, t_{id_n}\}\ t_{id_i} \in ClaT(cl_{id}) \end{cases}$$

- o : This identifies a set of operations that can be performed on the TH(s) to which the capability allows access. This is to note that, $o \subseteq O$ and,

  $$o = \{o_{id_1}, o_{id_2}, o_{id_3}, \ldots, o_{id_n}\} \mid \exists\, cl_{id} \in Cla \rightarrow o \subseteq ClaO(cl_{id})$$

  Note that the class ID (cl_{id}) here is the same class ID which we discussed above for t.
- Sig : This is the digital-signature of the issuer of the capability. This protects the integrity of the capability from being forged or tampered with.

- CoR : An optional set of condition rules that must be satisfied for access to be allowed. It is at the discretion of the THs how to interpret multiple rules (e.g. whether all must be satisfied or only one). Importantly, the CoR may reference local contexts. For example, the TH's location (e.g. a particular room in a building) or the date and time. Note, we are aiming for a minimal, but non-trivial, set of properties but other features can be added for domain specific applications.

Compared to the other capability structure e.g. [20], which follow a heavy-weight XML structure, we use JSON.

3.4.8 Capability Instantiation

Now we discuss how a capability is constructed by the capability instantiation process. The capability instantiation process generates an actual capacity from an appropriate capability template. A capability template can be seen as a 'frame' which specifies how a capability is constructed. Importantly, a capability template consists of the same fields as a capability, with the addition of the *Capability Parametrization Rule*. The value of some fields in the capability template will only be defined on actual capability generation. The Iss_{id}, t and o fields are fixed at the time of the creation of the capability template. The Cap_{id}, U_{id}, Iss_{time}, Exp_{time} and Sig fields are blank in the template and are filled in an actual capability at the time of capability generation. The CoR field will be partially defined in a capability template, its final 'value' is determined by the application of the $Capability\ Parametrization\ Rule$. In Fig. 3.8, we illustrate a conceptual construction of a simple capability template using JSON. Note, { } represents the place holders for the corresponding information which will be filled-up at the time of a capability generation and others are the actual values.

Now let us consider the above mentioned capability template of Fig. 3.8 to show how an actual capability is issued using appropriate *Capability Parameterization Rule*. As noted above, the encoding of information between different entities in our system can easily be achieved using JSON. The parameterization of a capability template to an actual capability will add some extra conditions on the THs to which the capability allows access by extending the CoR field that is held by the capability template. For instance, the class specified by the 't' field of the capability template could be the '$Lights\_in\_Building\_E6A$' and the capabilities produced from this template could be parameterized to only give access to lights on certain floors. In this case, the parameterization adds one extra 'rule' to the CoR. So, for example, the CoR could have added to it 'floor=floor_2' where all the lights in the building have an attribute 'floor'.

Another example would be a policy which allows doctors access to sensors attached to patients. Note that this example is related to the capability template illustrated in Fig. 3.8. Doctors are only to be allowed access to the patients of the ward to which they are assigned. So the 't' field of the capability template would hold

```
{
"Cap_id"   : {"~"},
"U_id"     : {"~"},
"Iss_id"   : "Hospital#H",
"Iss_time": {"~"},
"Exp_time": {"~"},
"t"    : "sensor",
"o"    : "read",
"Sig" : {"~"},
"CoR" : [{
         "~"
      }]
"CPR" : [{
         "user_attribute   : ward_assigned"
         "thing_attribute  : ward"
      }]
}
```

Fig. 3.8 A capability template. CPR denotes the *Capability Parametrization Rule*

the value 'sensor', a class to which all sensors attached to patients would belong. The parametrization rule would require the ward of assignment of the doctor. Assume a doctor is assigned to 'ward_2' and has a corresponding attribute. Then the CoR field of a capability issued to that doctor would include the rule 'ward=ward_2'. This assumes that all the sensors have an attribute 'ward' and that is set appropriately. Then the doctor in this example would only have access to sensors where the 'ward' attribute is set to 'ward_2'.

A parameterization rule consists of the name of an attribute that must be held by the user requesting access and the name of an attribute of the TH (or more properly class of TH) to which access is being requested. The CMS requests from the user the value of the nominated attribute. In the simplest form of the process the CMS forms the CoR field in the capability by taking the CoR field in the capability template and adding to it a new rule of the form:

$$'thing\_attribute' : val('user\_attribute')$$

In brief, such a rule ensures that TH has an attribute with the same value as that of the user. As the values of the attributes held by different users may differ, the capabilities produced for different users will have different rules and are thus parameterized. More than one attribute can be specified in the parameterization rule, allowing for more fine-grained access control. More generally, the rule could be of the form:

$$'thing\_attribute' = \\ f(val('user\_attribute'_1), ('user\_attribute'_2), \ldots, ('user\_attribute'_n)$$

That is, rather than simply copying the value of the user attribute in to the extension of the CoR field, a function can be used to create a value, which is the value that the attribute of TH must hold if access is to be granted using the generated capability.

Next, we give an example of how the aforementioned process happens using a capability template depicted in Fig. 3.8. Note, for simplicity, the capability template for this example holds no pre-set conditions in the CoR field.

Suppose, Doctors in Hospital 'H' are assigned to particular wards. Each Doctor will wish to access the output of sensors attached to the patients in their assigned ward. Each Doctor will have an attribute 'ward_assigned', which holds the value of the ward to which they are assigned. Each sensor will have an attribute 'ward'. This attribute holds the name of the ward to which the patient has been admitted. We assume that the appropriate capability template is assigned to a role and that the doctor in the following example has other attributes which prove their right to be a member of that role.

Assume Doctor A is assigned to 'ward_2' and wishes to obtain a capability giving them access to the sensors of patients admitted to 'ward_2'. When Doctor A first attempts to access such a sensor their user device will not hold an appropriate capability. After contacting the sensor, the Doctor A's device contacts the CMS. The CMS locates the appropriate role and capability template (permission) and notes that there is a capability parameterization rule for that template. When the CMS informs the user device of the attributes required for role membership, it also includes the user attributes included in the capability parameterization rule. The values of these attributes are then provided to the CMS by Doctor A's user device.

In the particular example from Fig. 3.8, for purposes of capability parameterization, the relevant attribute is 'ward_assigned'. In the case of Doctor A, the value of this attribute is 'ward_2'. As noted in the capability parameterization rule the TH has the attribute 'ward'. The CMS, in generating a capability from the capability template adds to the CoR field the condition:

$$ward : ward\_2$$

This condition means that, for access to be allowed, the TH must have the value of its 'ward' attribute be 'ward_2'. This is checked by the TH (or its manager) on access, as is the case for all other conditions in the CoR

The actual capability generated by the *Capability Parametrization Rule* is shown in Fig. 3.9. Note that if the user had a different value of the attribute 'ward_assigned' then the value of the CoR field would correspondingly vary. In other words, the capability would not allow a user to access to sensors with (for example) the attribute value pair 'ward : ward_3'.

Next, we provide a brief discussion of the capability instantiation process using a capability template in our proposed architecture.

In Fig. 3.10, we illustrate the capability instantiation process. When a request for a capability reaches the PMU via the RM (steps 1 and 2 of Fig. 3.10), the EE checks the corresponding access policies from the PD (step 3 of Fig. 3.10) and sends back the evaluation results to the EE (step 4 of Fig. 3.10). If at least one policy is satisfied, the EE contacts the CG (step 5a of Fig. 3.10). The CG then contacts the CD where capability templates are stored (step 6 of Fig. 3.10) to instantiate a capability from the appropriate capability template (step 7 of Fig. 3.10). Finally, the CG sends the

```
{
"Cap_id"   : "jXEPy0UFLzC4oa4ROYTCRTP39",
"U_id"     : "SN#12348484",
"Iss_id"   : "Hospital#H",
"Iss_time": "050619120000",
"Exp_time": "150619120000",
"t"    : "sensor",
"o"    : "read",
"Sig"  : "JhbGciECEF0OSQVMiLC0eXAPS",
"CoR"  : [{
        "ward : ward_2"
     }]
}
```


Fig. 3.9 An issued capability

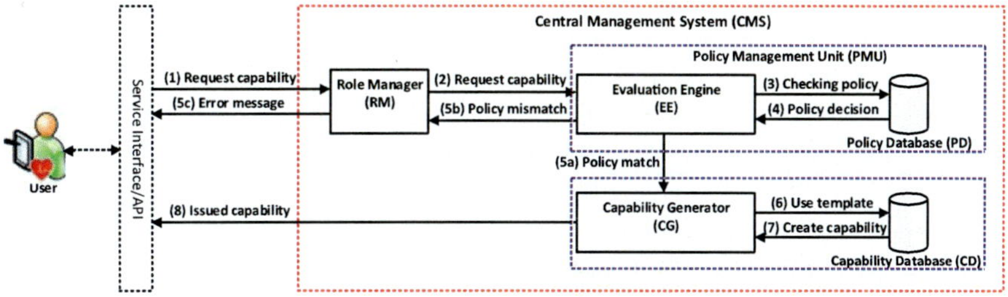

Fig. 3.10 The process of instantiating a capability from a capability template

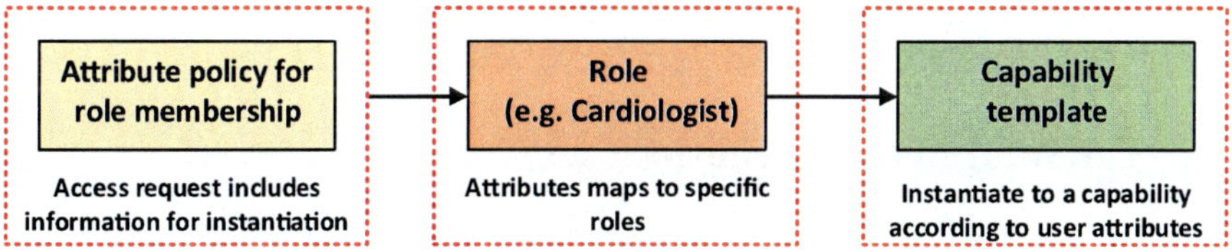

Fig. 3.11 From attributes to a capability instantiation

issued capability to the user (step 8 of Fig. 3.10). Note, if the corresponding policies do not match, then the request terminates, and a response is sent back to the user (steps 5b and 5c of Fig. 3.10).

In Fig. 3.11, we depict the use of attributes in role membership and capability instantiation. Recall, a capability template is composed of the fields needed to generate an actual capability. Some fields, e.g. the operations the capability provides access to (e.g. read, write, etc.) and conditions to be evaluated on capability use, may be pre-defined. Others, e.g. capability and user identity, expiry time and the exact THs the issued capability will allow access to, will be specified based on policies and other information stored in the system and attributes supplied by the requesting user. This means that different capabilities can result from the same capability template, even to the extent of allowing access to different THs.

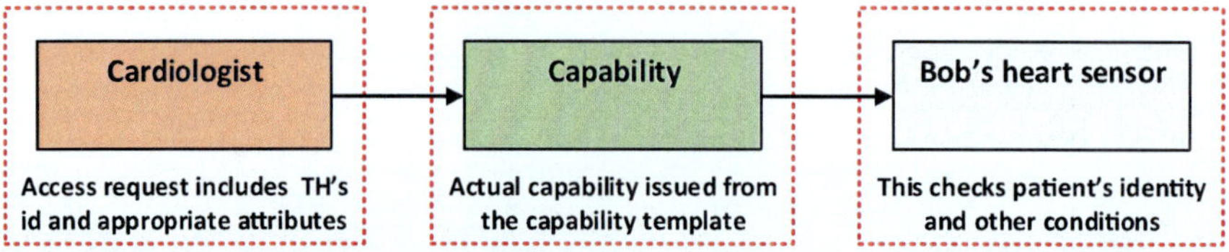

Fig. 3.12 Using a capability for accessing a TH

Capability templates may differ in how much variance they allow in instantiation. For example, a capability template may simply provide users with a capability for all doors of class *'Public_Access'*. All capabilities instantiated from such a template will grant access to the same set of resources. In other cases, a capability will provide more fine-grained access as defined by the parameterized rule. For example, referring to Fig. 3.3, a doctor with a role 'cardiologist' (e.g. Doctor A) wants to access the heart sensor of a patient (e.g. Bob). In this case, Doctor A sends a request along with attributes satisfying role membership and attesting to their status as the 'cardiologist of Bob'. The capability template will state that it allows access to the THs of class *'heart_sensor'* attached to a patient for whom the user is the doctor. The instantiated capabilities will then only include the identity of heart sensors registered to patients for whom the requester has provided attributes attesting that they are the cardiologist of that patient. In this way multiple access situations can be governed by a single policy expression. In Fig. 3.12, we depict the use of a capability in obtaining access of a TH.

3.5 Different Access Scenarios

In this section, we explore different potential access scenarios for our proposed access control architecture. We return to the use case example that we discussed in Sect. 3.3, and outline different access scenarios based on the issued capability, different access operations on THs and various conditions associated with an access.

Scenario 1—First Access: In this scenario, a user (i.e. the UD in our architecture) receives APIs from a TH. The UD communicates with the CMS requesting a specific service from a specific TH. The UD needs a capability to perform an operation and we assume that the UD does not have an appropriate capability. The UD needs to send appropriate attributes to the CMS to satisfy the role-membership. If satisfied, the CMS issues the capability. The UD requests access to the operation from the TH and presents the capability. The TH checks that the capability authorizes the requested access, via the algorithm (either Algorithm 3.1 or Algorithm 3.2) that we discussed below. If the algorithm returns 'granted' the UD is allowed access. For example, Doctor A can access Bob's clinical sensors with a valid capability.

Scenario 2—Subsequent Accesses, Same *'Thing'*, Same Operation: In this scenario, a UD wishes to repeat an operation on a TH for which the user has already obtained an appropriate capability. As the UD already has an appropriate capability it makes the access request directly to the TH, presenting the capability. The TH again checks that the capability authorizes the requested access, via appropriate algorithm. If the algorithm returns 'granted' the UD is allowed access. Note that the CMS is not involved in this scenario and that the UD did not need to obtain a new capability. For example, Doctor A can access Bob's cardiac sensors several times after obtaining a capability without consulting the CMS after the first access.

Scenario 3—Subsequent Accesses, Same *'Thing'*, Different Operation: Capabilities may allow access to multiple operations, and such capabilities can be used to access operations other than that for which the capability was initially requested. If a capability that the UD holds allows the access, refer to scenario 2.

Scenario 4—Access to Multiple *'Things'* with a Single Capability: Capabilities may allow access to multiple THs. With the first access, the capability is obtained as in scenario 1. For subsequent accesses the UD contacts the new TH, identifies that it already holds an appropriate capability by searching the database of capabilities stored on it. It then presents the capability along with the access request as in scenario 2. For example, a nurse is allowed to access the body temperature and blood pressure sensors of multiple patients (e.g. Bob and Alice) using a single capability. Note that if the capability allows access to multiple THs and multiple operations on those THs, then access to a different TH may involve a different operation to the initial access.

Scenario 5—Invalid Issuer of the Capability and/or Signature on Request: This scenario, in particular, important for an asymmetric key based approach (discussed in Sect. 3.7). A UD has a capability and wants to perform a desired operation. However the capability has not been provided by an issuer (in this case the CMS) that the TH recognizes. The UD presents the capability to the TH along with the access request. When the TH checks the signatures on the capability and the request it will reject the request (Algorithm 3.2 returns 'refused') and the access will not be allowed.

Scenario 6—Capability has Expired: A UD has a capability that allows access to certain THs but the capability has expired. If the UD detects this, then refer to scenario 1. If the UD presents it to the TH anyway then TH checks the capability and discovers that the time of expiration of the capability has been reached. Either Algorithm returns 'refused'. If the UD wishes to obtain access they need to request a new capability for the particular access required, which may be obtained as per scenario 1.

Scenario 7—Validating Local Conditions: A capability may contain condition rules which must be validated by the TH before access is granted. Conditions can involve context e.g. correct date and time, location, etc. or properties of the TH itself, e.g. available storage, remaining battery power and any other conditions related to the state of the TH itself. Thus, when the UD sends a capability to the TH, along with all the checks mentioned above (cf. scenario 1) the TH checks the '$Condition\ Rule$' in the capability. If the $Condition\ Rule$ (i.e. the CoR) are successfully validated, the access is allowed (in this case either Algorithm returns 'granted') otherwise access is refused.

3.6 System Operation: Symmetric Key Approach

In this section, we describe one implementation of our design, employing a symmetric key approach. This will help us to examine the suitability of light-weight cryptographic techniques for resource constrained IoT devices. In Fig. 3.13, we illustrate the design of the system using symmetric key based approach.

Note, the users and the THs must be registered in the system. Access to a TH begins when a UD detects a service that its user wishes to access. Assuming that the UD does not already hold an appropriate capability, this is followed by a request for an appropriate capability and then by access to the requested resource. Figure 3.14, shows an overview of the communication protocol for an access request.

3.6.1 *Registration*

Any new user and THs must be registered with the system. On registration they are provided with long-term symmetric-keys for communication with the central part of the system. Recall, the symmetric key based approach uses cryptographic algorithms that use the same keys for both encryption of plain text and the decryption of cipher text. The encryption key can be created and stored on a key management server. In our architecture, this could be situated inside the CMS. These keys can be distributed mutually or automatically between two parties by the key management server. The details of this are outside the scope this book but would follow the normal processes

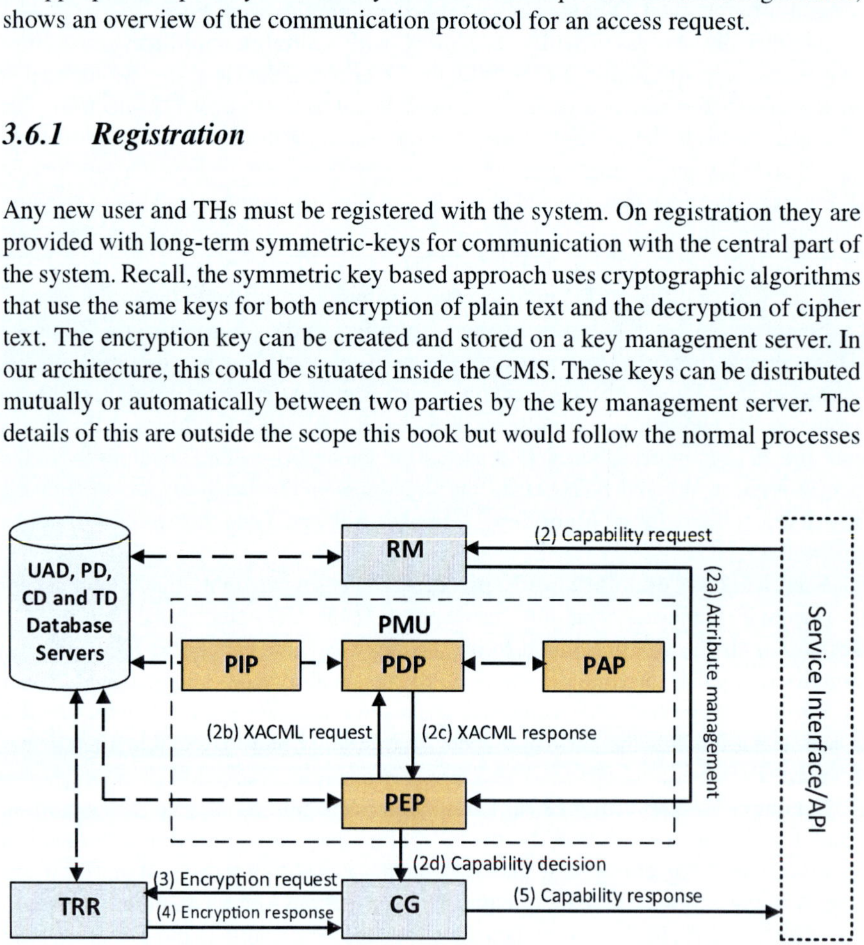

Fig. 3.13 The functioning of the system using a symmetric key based approach. Note, the steps discussed here represents the same communication that are depicted in Fig. 3.14, however, in more detail. The intermediate steps 2a, 2b, 2c and 2d depict the same functions as discussed in Fig. 3.6 (i.e. steps a, b, c and d)

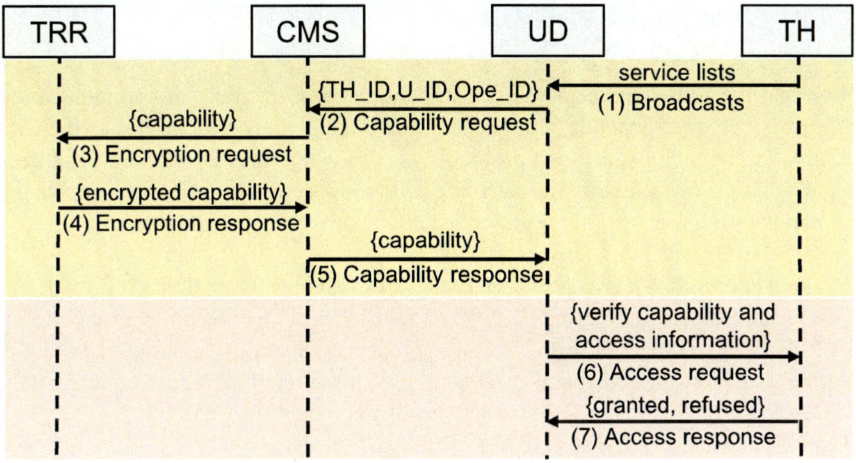

Fig. 3.14 The communication associated with a symmetric key based approach

for registration. Attributes of both users and THs are stored. For users, storage is in the UAD, for THs, it is in the TRR. User attributes include name, age, role membership, etc. TH attributes may include location, function, patient or ward assignment, etc. Attributes that apply to a registered user may be supplied by that user, assigned by the CMS or obtained from a trusted third party. Policies are written based on these attributes, not on concrete user identity. Users are provided with a password which is hashed and securely stored by the CMS.[3] This password hash is then used to generate a secret key K_{UD} to encrypt a message containing the session key to be used for the communication between the UD and the CMS. The ability to decrypt this message authenticates the UD to the CMS and ultimately prove its identity to the TH. As part of a TH's registration the TRR generates a secret key K_{TH} which is stored locally. That secret key is given to the TH. Thus, the registration process creates a bidirectional trust between the TRR and the TH. Group keys may also be provided to the TH. For example, the THs governing the lights on a given building floor may share a group key for communication with the CMS.

As K_{UD}, K_{TH} and possible group keys are long-term keys they should be used sparingly. Shorter term keys may be periodically refreshed using the long-term keys and employed for actual communication between the central system and registered entities. However, such key management issues are outside the scope of this chapter and in the following K_{UD} and K_{TH} are used for convenience.

[3]Note that another hash of this password may be stored on some other part of the system, e.g. an LDAP server, for supplementary service access.

3.6.2 Generating a Capability

Note that the following steps may be implemented in a variety of communication protocols; we have generally omitted the sender and receiver identities, as they are assumed to be included in such protocols, except where they are needed to address security issues or for clarity. As such the following steps are meant to illustrate the implementation flow and not represent a fully detailed, secure, protocol.

• TH → Broadcast: $TH_{ID}, \{Ope_{ID}\}$

The TH broadcasts its identity and the set of services (operations) it provides to UDs located in proximity using IEEE 804.15.4 BLE beacon or a similar protocol (step 1 of Fig. 3.14). If the UD possesses an appropriate capability (assume that the UD already holds one valid capability), then the communication proceeds immediately to step 6 of Fig. 3.14, otherwise the UD needs to contact the CMS.

• UD → CMS: $U_{ID}, \{U_{ID}, TH_{ID}, Ope_{ID}, T_{UD}, N_{UD}\}_{K_{UD}}$

When a user wishes to access a service, the UD requests a capability from the CMS by sending the identity of the user, the TH and the operation required (step 2 of Fig. 3.14). This is encrypted under K_{UD}/CMS, the key the UD shares with the CMS to prevent eavesdroppers identifying which service is being requested. A timestamp (T_{UD}) and nonce (N_{UD}) are also included for freshness purposes.

• CMS → UD: $\{Cap, K_{UD,TH}, T_{UD}, N_{UD}, \{Cap, K_{UD,TH}, T_{TH}, N_{TH}\}_{K_{TH}}\}_{K_{UD}}$

Upon receiving the request, the CMS uses the RM to determine what role(s) give access to the TH and what attributes must be supplied for role membership. The UAD supplies the attributes associated with the user. If the UAD supplied attributes match the attribute expression for role membership, the requested capability (denoted as Cap) is instantiated and generated by the CG, along with a session key for use between the UD and the TH. Then the CMS contacts the TRR for a long-term key associated with the TH so that the capability can be encrypted (step 3 of Fig. 3.14). The TRR performs the necessary encryption and sends the encrypted capability to the CMS (step 4 of Fig. 3.14). The communication between the CMS and the TRR must be performed over a secure medium and preferably using end-to-end encryption.[4]

The CMS then sends the capability and the session key ($K_{UD,TH}$), encrypted under a key shared with the UD, to the UD (step 5 of Fig. 3.14). This information is also encrypted under the key the TH shares with the TRR (K_{TH}/TRR). The latter will enable the TH to check the validity of any request. The capability includes the identity of the TH as one of its fields, enabling the UD to determine which request to the CMS this was a response to.

[4]Public key cryptography is an option here as the CMS and TRR are hosted on non-resource-constrained hardware. Note, in Sect. 3.7, we provide a public key based approach for implementation to show the differences in the performances.

3.6.3 Processing an Access Request

After receiving the capability from the CMS, the UD may now send an access request to the TH (step 6 of Fig. 3.14). The communication between the UD and the TH represents the most resource-constrained aspects of the system, because of the potentially limited capacities of the THs. Recall, in our proposed access control architecture a capability is checked by the edge devices upon access.

- UD →TH: $\{Cap, K_{UD,TH}, T_{TH}, N_{TH}\}_{K_{TH}}, \{U_{ID}, Ope_{ID}, T'_{UD}, N'_{UD}\}_{K_{UD,TH}}$
An access request is composed of the following two pieces of encrypted data:

1. *Capability*: This remains encrypted using the key K_{TH} and contains the permissions assigned to the user.
2. *Access Token*: Created by the user. It is encrypted using the session key $K_{UD,TH}$. It also contains: $\{U_{ID}, Ope_{ID}, T'_{UD}, N'_{UD}\}$. A new timestamp ($T'_{UD}$) and nonce ($N'_{UD}$) are included to help protect against replay attacks.

Upon receiving this message, the TH can decrypt the capability using the key shared with the TRR. It can then obtain the session key and decrypt the remainder of the message. Finally, the TH checks the capability as outlined in Algorithm 3.1.

Algorithm 3.1 Capability authorization process (a symmetric key approach)

receive(encCapability, encAccessToken)
capability, $K_{\mathrm{UD,TH}}$ ← **decrypt**(encCapability, K_{TH})
access ← **decrypt**(encAccessToken, $K_{\mathrm{UD,TH}}$)
if capability.ThingID = this.ID **and** access.ThingID = this.ID
  and access.UserID = capability.UserID
  and access.OperationID **in** capability.OperationID
  and access.Condition ⊂ capability.Condition
  and capability.ValidFrom ≤ access.Timestamp ≤ capability.ValitUntil
  then
    result ← *'granted'*
else
    result ← *'refused'*
end
send(**encrypt**(result, $K_{\mathrm{UD,TH}}$))

The TH checks a number of conditions that are listed in the capability. It ensures that the current time is within the period defined by the issued and expiry fields of the capability. It also verifies that the requested TH's ID matches with the specific TH's identity to which the access request is made. The user ID also verifies by the TH to ensure that the issued capability is granted to the specific user. The operation ID on the capability need to be the same that of the access request. The TH also verifies that the any other conditions contained within the capability are satisfied. Conditions can involve context e.g. correct date and time, the location, etc., or properties of the TH itself, e.g. available storage, remaining battery power and any other conditions related to the state of the TH itself. These conditions are listed in a capability and are evaluated locally within the THs. Algorithm 3.1 returns a decision on whether the access request is granted or refused. The TH generates the access decision (step 7 of Fig. 3.14).

As only the UD to whom the capability was issued and the TH knows the key $K_{UD,TH}$, only that UD would be able to formulate the message. Intuitively, the capability specifies the summary of policy rules applicable to the TH and the UD given the context while the access token specifies the actual action that the user wants to perform on a specific TH.

3.7 System Operation: Asymmetric Key Approach

In this section, we describe another implementation of our design, employing asymmetric key approach. In general, symmetric key encryption algorithms are much faster in computation and therefore need relatively less computational power than asymmetric encryption algorithms, but their main weakness is key distribution. Especially, the symmetric key based approach is useful to apply when speed and computing power are the primary concerns [55]. We argue that a system that is built upon a symmetric key based approach can easily be built based on the asymmetric key based approach.

Note, the architectural components (cf. Sect. 3.4.4) of the proposed access control architecture remain the same. The users need to supply required attributes at the time of an access request made to the CMS. In Fig. 3.15, we illustrate the protocol for satisfying a user request using an asymmetric key based approach. Note, unlike Fig 3.14, in this case, the TRR is not involved in the communication when generating a capability.

3.7.1 *Registration*

The THs must be registered with the central part of the system. Note, in this case, the TH's registration is only required for authorization purposes. The authorization decisions are handled by the access control policies. In the previous approach (i.e. systemic key based approach), the TH's registration is required for both the authentication and authorization. Authentication is done by sharing the keys in a secret way. However, in an asymmetric key based approach, this is avoidable as this is performed by the public key of the CMS. Recall, in an asymmetric key based approach (also known as the public key cryptography) two different keys are employed for encryption and deception. The public key is used to encrypt the plain text and the private key is used to decrypt the cipher text. Fundamentally, these keys are simply large numbers that have been paired together using specific cryptographic algorithms but are not identical. For sharing the public key, secure SSH connection can be used.

Public key can be freely shared with everyone, while the private key must need to be kept secret. Once again, the details of the generation of keys is outside the scope of this book but would follow the normal processes for registration. User's attributes are stored in the UAD or some other location accessible by the PIP and for the THs, it is stored in the TRR. User attributes include name, age, role membership, etc. TH's attributes may include location, function, patient or ward assignment, etc. In this case, attributes that apply to a role membership must be supplied by the corresponding user. Recall, in our proposal, policies are written based on the attributes not on concrete user identity.

3.7.2 Generating a Capability

Once again, note that the following steps are meant to illustrate the implementation flow. This should not be seen as the representation of a fully detailed, secure, protocol.

• TH → Broadcast: TH_{ID}, $\{Ope_{ID}\}$

Like the previous case (cf. Sect. 3.6.2), here we also assume that the TH broadcasts the services (operations) it provides to UDs located in proximity using IEEE 802.15.4 BLE beacon or a similar protocol (step 1 of Fig. 3.15). If the UD possesses an appropriate capability, then the communication proceeds to step 6 (of Fig. 3.15). Otherwise, the UD communicates with the CMS, specifying the TH and service it wishes to access, in order to obtain an appropriate capability.

• UD → CMS: $\{TH_{ID}, Ope_{ID}, T_{UD}, N_{UD}\}$

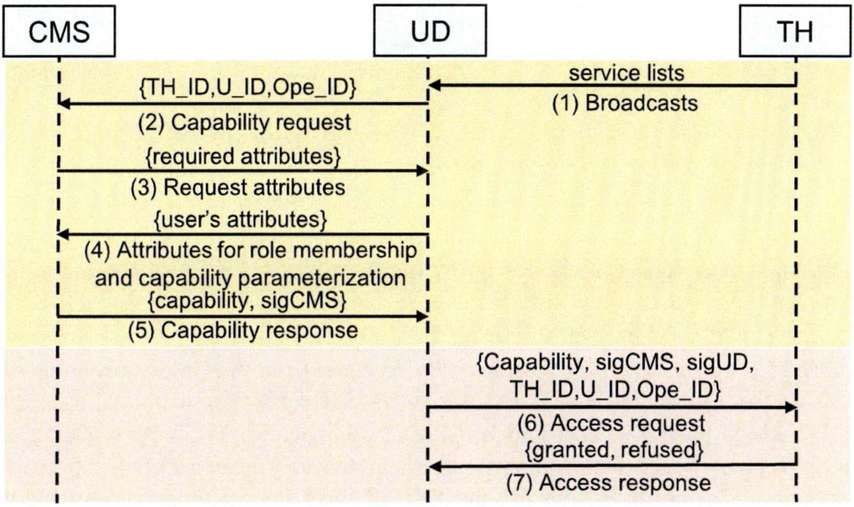

Fig. 3.15 The communication associated with an asymmetric key based approach

The UD requests a capability from the CMS by sending the identity of the TH and the operation (Ope) required (step 2 of Fig. 3.15). A timestamp (T_{UD}) and nonce (N_{UD}) are also included for freshness purposes. We assume that all of the communications between the CMS and the UD are performed over a secure medium and preferably using end-to-end encryption.

• CMS → UD: $\{Attr, U_{ID}\}$

Upon receiving the request from the UD, the CMS contacts to the RM to determine what role(s) give access to the TH and what attributes ($Attr$) the user must supply for role membership. Then the CMS contacts to the UD with the required attributes. In particular, the CMS uses the RM and CD to locate the appropriate capability template and extracts the necessary rules from the PD. This is used to inform the particular UD of the attributes that must be presented (step 3 of Fig. 3.15).

• UD → CMS: $\{Attr, U_{ID}\}$

These attributes are those required to obtain role memberships and (if specified) further attributes required for capability parameterization. Assuming that the UD holds, on behalf of the user, attributes that will satisfy the requirements it sends them, and the user identity, to the CMS (step 4 of Fig. 3.15).

• CMS → UD: $\{Cap, CMS_{Sig}\}$

Upon receiving the required attributes, the CMS checks that the supplied attributes satisfy the requirements for role membership. The CMS can also get some extra attributes from other sources, if required (e.g. from the PIP). It then creates a capability (Cap) for the requested *thing*/operation pair, by filling in the capability template with the user's identity, any necessary time stamps (e.g. beginning and end times for capability lifespan) and any parameterization information. The user's identity is required to ensure that the capability is not passed to an unauthorized user. At this stage, the capability is encrypted with the signature of the CMS (CMS_{Sig}—i.e. the public key of the CMS), and is then sent to the UD from where the specific access request has made (step 5 of Fig. 3.15).

3.7.3 *Processing an Access Request*

The UD may now present the capability to the TH for accessing a specific resource with the issued capability. Recall, this is the most resource-constrained aspects of the system.

• UD → TH: $\{\{Cap, CMS_{Sig}\}_{UD_{Sig}}, U_{ID}, Ope_{ID}, T'_{UD}, N'_{UD}\}$

The UD further encrypts the capability received from the CMS by signing the request, i.e. using UD's signature(UD_{Sig}—i.e. the public key of the UD), to the TH (step 6 of Fig. 3.15). This is important because it helps to verify the sender of an access request. Recall, the CMS's signature helps to verify the issuer of the capability. Upon receiving the access request, the TH will check the capability, as outlined in Algorithm 3.2, including checks on the CMS's signature on the capability (using the private key of the CMS) and the UD's signature on the request (using the private key of the UD) and reply to the UD (step 7 of Fig. 3.15). Note, the TH needs to decrypt two signatures associated with the capability, one from the CMS and the other from the UD. The access request also contains U_{ID}, Ope_{ID}, T'_{UD} and N'_{UD}. A new timestamp (T'_{UD}) and nonce (N'_{UD}) are included to help protect against replay attacks.

Algorithm 3.2 Capability authorization process (an asymmetric key approach)

1: **receive**(encCapability, encAccessToken)
2: capability, ← **decrypt**(encCapability, Sig_{UD})
3: access ← **decrypt**(encAccessToken, Sig_{CMS})
4: Decision ← *'refused'*
5: **if** capability is not Null **then**
6: **if** valid (time stamp) **then**
7: **if** $user_{id} = capability_{user}$ **then**
8: **if** TH_{id} in $capability_{TH}$ **then**
9: **if** ope_{req} in $capability_{ope}$ **then**
10: **if** CoR = true **then**
11: **if** valid UD_{sig} and valid CMS_{sig} **then**
12: Decision ← *'granted'*

Algorithm 3.2 takes the capability supplied by the user, the operation requested, the user's identity, the identity of the TH and the signatures on the request (of the UD) and capability (of the CMS). It checks that the current time is within the period defined by the issued and expiry fields of the capability, that the user making the request was the one to whom the capability was granted, the capability allows access to the requested TH and operation, that any condition rules contained within the capability are satisfied and that the signatures are valid. Signature checks are left to last as they are the most-consuming operation. Conditions can involve context e.g. correct date and time, the location, etc. or properties of the TH itself, e.g. available storage, remaining battery power and any other conditions related to the state of the TH itself. These conditions are listed in a capability and are evaluated locally within the THs. The algorithm returns a decision on whether the requested operation is granted or refused.

3.8 Discussion

In the foregoing discussion we have focused on a single CMS and how it governs access to IoT devices registered with it. In the real-world, a single CMS would not be practical for the entirety of the IoT. The IoT will consist of multiple administrative domains, each with their own policies, resources and THs. Our design can easily be extended to a multiple domain situation. Each administrative domain would have its own CMS, which would hold the access control policies for that domain and with which the THs in that domain would register. More precisely, the THs would register with the TRR associated with the CMS of that domain and be governed by the access control policies stored in the CMS. The CMS would issues capabilities for THs registered with its associated TRR. A domain then consists of the THs registered with the TRR and the CMS of the domain. As policies are held within the CMS, each domain can have its own policies.

Users are not required to register within a particular domain. The CMS of a domain would recognize certain attribute providers. To access THs in the domain of the CMS users would need to hold attributes supplied by attribute providers accepted by the CMS of the domain. For example, as illustrated in Fig. 3.16, the UD wishes to access the TH. The UD receives attributes from the attribute provider (labelled as 'AP'). These attributes are recognized and accepted by the CMS of the domain and, assuming all other conditions are met, access is allowed. Note that the CMS may not have been aware of the existence of the UD until access is requested. The CMS will

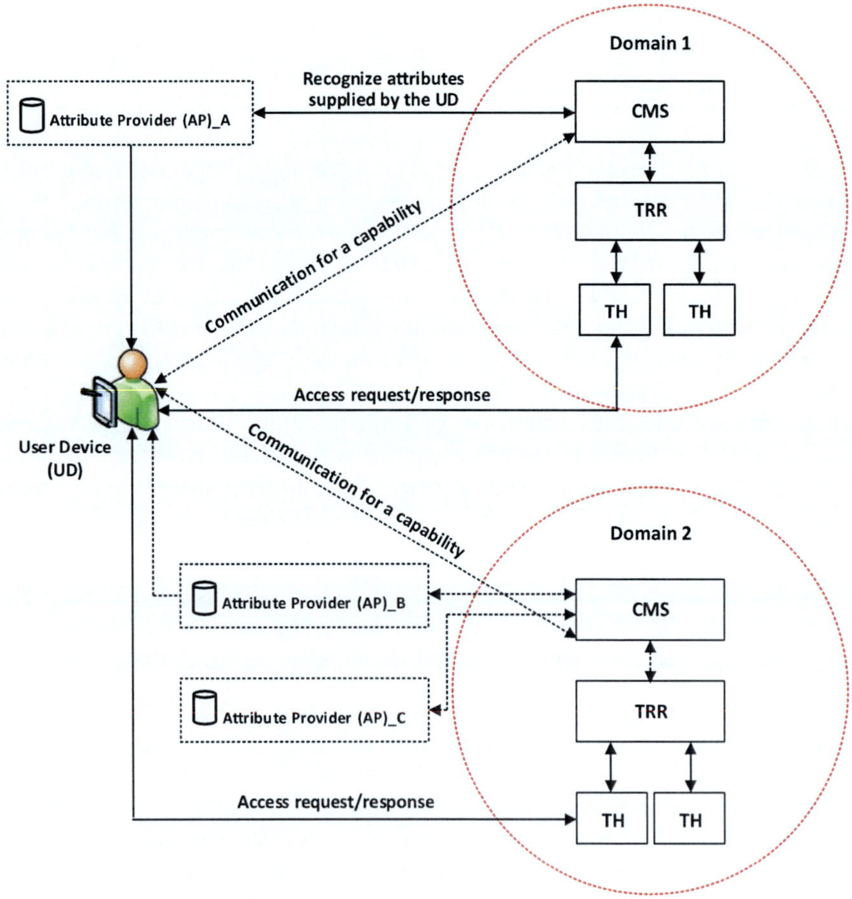

Fig. 3.16 A multiple domain scenario

have been aware of the attribute provider, and what attributes it can provide, but not to what users they are provided. A CMS may accept attributes supplied by multiple attribute providers. The sets of attributes supplied by these attribute providers may or may not overlap. Which attribute providers are acceptable to a CMS is determined entirely by the administrative policies of that domain.

A user may have the same attribute attested by multiple attribute providers. This is necessary as the CMS of different domains may recognize different attribute providers but reference the same attribute in their policies. In our design, it does not matter which attribute provider supplies a user with an attribute, merely, that the user has the attribute supplied by a recognized attribute provider. A CMS can advertise, as part of its service, which attribute providers it recognizes. The UD can then supply the appropriate attributes, attested by a recognized attribute provider. Users can then interact with any domain, providing they have attributes supplied by an attribute provider recognized by that domain. Capabilities will only be valid in the domain in which they are issued, as the THs will only accept capabilities from the CMS of their domain.

Note that the CMSs of different domains do not need to communicate with each other, as they are responsible solely for access within their own domain. A CMS is free to choose which attribute providers it will recognize, allowing them to accept attributes only from trusted attribute providers. While there is some burden imposed on UDs in holding attributes from multiple providers, this will be lessened if attributes providers are recognized by multiple CMSs.

3.9 Summary

In this chapter, we have outlined the design of a general access control system for the IoT that combines elements of attributes, roles and capabilities to achieve streamlined policy management. This provides a flexible framework on which to build an access control architecture in a smart environment. Our design employs attributes for role management, capabilities for access right implementation, and attributes for fine-grained policy decisions based on capabilities. We also propose a partially decentralized architecture for real-time decision making, which can help to achieve better performance for IoT systems using light-weight security mechanisms. Access rights are embodied in capabilities. The capabilities are provided to users on request, based on the attributes of users and the roles which those users give them membership of. Once a capability is obtained, the user attributes do not need to be checked again while the capability remains valid. The edge devices (i.e. the *things*) need only to check the capability, avoiding any communication with a central system at that point, including any need for repeated attribute evaluations. To minimize the policy expressions required the capabilities issued to users as a result of a single policy may differ, based on the attributes possessed both by the user and the target devices.

With a number of use cases, we showed the practicability of proposed access control solutions that can efficiently provide the required policy management for different situations. In our proposed access control model, role membership is specified using attributes, not explicit user to role assignment. This provides a greater level of flexibility and conciseness in role management as updates to individual role membership do not have to be noted. It also allows a consistent and easily managed approach to which users will be members of which roles. By employing roles, rather than a direct mapping between attribute expressions and permissions, we do not have to needlessly duplicate attribute expressions. It also allows us to take advantage of the power of the role-hierarchies of RBAC. We used capabilities as credentials to govern access to *things*. Capabilities are distributed to users and presented to the edge devices (i.e. *things*) for access. This reduces the centralization of the system and eliminates a potential performance bottleneck.

In the next chapter (Chap. 4), we will demonstrate the feasibility of the proposed architecture with a practical proof of concept prototype implementation. As we discussed, our intention would be to employ both the light-weight key management protocol (i.e. symmetric key cryptography) and the PKI-based approaches (i.e. asymmetric key cryptography) for authentication and secure communication. We also intend to compare our findings with existing approaches and to provide a comprehensive adversary analysis. In addition, we aim at providing a comparison of number of policy expressions required in our case with the available approaches.

References

1. Z. Yan, P. Zhang, A.V. Vasilakos, A survey on trust management for internet of things. J. Netw. Comput. Appl. **42**, 120–134 (2014), http://www.sciencedirect.com/science/article/pii/S1084804514000575
2. D. Evans, The Internet of Things: How the Next Evolution of the Internet Is Changing Everything (2011), https://www.cisco.com/c/dam/en_us/about/ac79/docs/innov/IoT_IBSG_0411FINAL.pdf. Accessed 10 Oct 2017
3. B. Farahani, F. Firouzi, V. Chang, M. Badaroglu, N. Constant, K. Mankodiya, Towards fog-driven iot ehealth: promises and challenges of iot in medicine and healthcare. Future Gen. Comput. Syst. **78**, 659–676 (2018), http://www.sciencedirect.com/science/article/pii/S0167739X17307677
4. S.M. Riazul Islam, D. Kwak, M. Humaun Kabir, M. Hossain, K.-S. Kwak, The internet of things for health care: a comprehensive survey. IEEE Access **3**, 678–708 (2015), http://dx.doi.org/10.1109/access.2015.2437951
5. NHS Test Bed, https://www.england.nhs.uk/ourwork/innovation/test-beds/. Accessed 30 May 2018
6. QardioCore, https://www.getqardio.com/. Accessed 21 May 2018
7. Zanthion, http://www.zanthion.com/. Accessed 20 May 2018
8. M. Dabbagh, A. Rayes, Internet of things security and privacy, in *Internet of Things From Hype to Reality* (Springer International Publishing, Berlin, 2017), pp. 195–223, http://dx.doi.org/10.1007/978-3-319-44860-2_8
9. M. Burhan, R.A. Rehman, B. Khan, B.-S. Kim, Iot elements, layered architectures and security issues: a comprehensive survey. Sensors **18**(9) (2018), http://www.mdpi.com/1424-8220/18/9/2796
10. M. Conti, A. Dehghantanha, K. Franke, S. Watson, Internet of things security and forensics: challenges and opportunities. Future Gen. Comput. Syst. **78**, 544–546 (2018), http://www.sciencedirect.com/science/article/pii/S0167739X17316667

11. T. Mahler, N. Nissim, E. Shalom, I. Goldenberg, G. Hassman, A. Makori, T. Kochav, U. Elovici, Y. Shahar, Know Your Enemy: Characteristics of Cyber-Attacks on Medical Imaging Devices (2018). arXiv:1801.05583
12. W. Sun, Z. Cai, Y. Li, F. Liu, S. Fang, G. Wang, Security and privacy in the medical internet of things: a review. Secur. Commun. Netw. **2018** (2018). https://doi.org/10.1155/2018/5978636
13. WIRED, How the Internet of Things got Hacked (2015), https://www.wired.com/2015/12/2015-the-year-the-internet-of-things-got-hacked/. Accessed 01 Oct 2017
14. Y. Lu, L.D. Xu, Internet of things (iot) cybersecurity research: a review of current research topics. IEEE Int. Things J. 1–1, (2018). https://doi.org/10.1109/JIOT.2018.2869847
15. L. Atzori, A. Iera, G. Morabito, The internet of things: a survey. Comput. Netw. **54**(15), 2787–2805 (2010), http://dx.doi.org/10.1016/j.comnet.2010.05.010
16. A. Solanas, F. Casino, E. Batista, R. Rallo, Trends and challenges in smart healthcare research: a journey from data to wisdom, in *2017 IEEE 3rd International Forum on Research and Technologies for Society and Industry (RTSI)*. (IEEE, 2017), pp. 1–6, http://dx.doi.org/10.1109/rtsi.2017.8065986
17. L.M. Tarouco, L.M. Bertholdo, L.Z. Granville, L.M. Arbiza, F. Carbone, M. Marotta, J.J. de Santanna, Internet of things in healthcare: interoperatibility and security issues, in *2012 IEEE International Conference on Communications (ICC)* (IEEE, 2012), pp. 6121–6125, http://dx.doi.org/10.1109/icc.2012.6364830
18. D.F. Ferraiolo, R. Sandhu, S. Gavrila, D.R. Kuhn, R. Chandramouli, Proposed NIST standard for role-based access control. ACM Trans. Inf. Syst. Secur. **4**(3), 224–274 (2001). http://dx.doi.org/10.1145/501978.501980
19. E. Yuan, J. Tong, Attributed Based Access Control (ABAC) for web services, in *Proceedings of the IEEE International Conference on Web Services*, ser. ICWS '05 (IEEE Computer Society, Washington, DC, USA, 2005), pp. 561–569, http://dx.doi.org/10.1109/icws.2005.25
20. S. Gusmeroli, S. Piccione, D. Rotondi, A capability-based security approach to manage access control in the Internet of Things. Math. Comput. Model. **58**(5–6), 1189–1205 (2013), http://dx.doi.org/10.1016/j.mcm.2013.02.006
21. Y. Zhang, X. Wu, Access control in internet of things: a survey (2016). arxiv:1610.01065.pdf
22. A. Ouaddah, H. Mousannif, A. Abou Elkalam, A. Ait Ouahman, Access control in the internet of things: big challenges and new opportunities. Comput Netw. **112**, 237–262 (2017). http://dx.doi.org/10.1016/j.comnet.2016.11.007
23. J.L. Hernández-Ramos, A.J. Jara, L. Marín, A.F. Skarmeta Gómez, DCapBAC: embedding authorization logic into smart things through ECC optimizations. Int. J. Comput. Math. **93**(2), 345–366 (2016), http://dx.doi.org/10.1080/00207160.2014.915316
24. J. Hernandez-Ramos, A. Jara, L. Marın, A. Skarmeta, Distributed capability-based access control for the internet of things. J. Int. Serv. Inf. Sec. **3**(3/4), 1–16 (2013), http://isyou.info/jisis/vol3/no34/jisis-2013-vol3-no34-01.pdf
25. S.B. Baker, W. Xiang, I. Atkinson, Internet of things for smart healthcare: technologies, challenges, and opportunities. IEEE Access **5**, 26 521–26 544 (2017). https://doi.org/10.1109/ACCESS.2017.2775180
26. P. Gope, T. Hwang, Bsn-care: a secure iot-based modern healthcare system using body sensor network. IEEE Sensors J. **16**(5), 1368–1376 (2016). https://doi.org/10.1109/JSEN.2015.2502401
27. C.A. Tokognon, B. Gao, G.Y. Tian, Y. Yan, Structural health monitoring framework based on internet of things: a survey. IEEE Int. Things J. **4**(3), 619–635 (2017). https://doi.org/10.1109/JIOT.2017.2664072
28. F. Firouzi, A.M. Rahmani, K. Mankodiya, M. Badaroglu, G. Merrett, P. Wong, B. Farahani, Internet-of-things and big data for smarter healthcare: from device to architecture, applications and analytics. Future Gen. Comput. Syst. **78**, 583–586 (2018), http://www.sciencedirect.com/science/article/pii/S0167739X17319726
29. M.A. Sahi, H. Abbas, K. Saleem, X. Yang, A. Derhab, M.A. Orgun, W. Iqbal, I. Rashid, A. Yaseen, Privacy preservation in e-healthcare environments: state of the art and future directions. IEEE Access **6**, 464–478 (2018). https://doi.org/10.1109/ACCESS.2017.2767561

30. L.Y. Yeh, P.Y. Chiang, Y.L. Tsai, J.L. Huang, Cloud-based fine-grained health information access control framework for lightweight iot devices with dynamic auditing and attribute revocation. IEEE Trans. Cloud Comput. 1–1, (2015). https://doi.org/10.1109/TCC.2015.2485199
31. D.A. Gandhi, P.M. Ghosal, Intelligent healthcare using iot:a extensive survey, in *2018 Second International Conference on Inventive Communication and Computational Technologies (ICICCT)* (2018), pp. 800–802, http://dx.doi.org/10.1109/ICICCT.2018.8473026
32. G. Aceto, V. Persico, A. Pescap, The role of information and communication technologies in healthcare: taxonomies, perspectives, and challenges. J. Netw. Comput. Appl. **107**, 125–154 (2018), http://www.sciencedirect.com/science/article/pii/S1084804518300456
33. Y. Yin, Y. Zeng, X. Chen, Y. Fan, The internet of things in healthcare: an overview. J. Ind. Inf. Integr. **1**, 3–13 (2016), http://www.sciencedirect.com/science/article/pii/S2452414X16000066
34. P. Dineshkumar, R. SenthilKumar, K. Sujatha, R.S. Ponmagal, V.N. Rajavarman, Big data analytics of iot based health care monitoring system, in *2016 IEEE Uttar Pradesh Section International Conference on Electrical, Computer and Electronics Engineering (UPCON)* (2016), pp. 55–60, http://dx.doi.org/10.1109/UPCON.2016.7894624
35. B. Xu, L.D. Xu, H. Cai, C. Xie, J. Hu, F. Bu, Ubiquitous data accessing method in iot-based information system for emergency medical services. IEEE Trans. Ind. Inf. **10**(2), 1578–1586 (2014). http://dx.doi.org/10.1109/TII.2014.2306382
36. R.M. Gomathi, G.H.S. Krishna, E. Brumancia, Y.M. Dhas, A survey on iot technologies, evolution and architecture, in *2018 International Conference on Computer, Communication, and Signal Processing (ICCCSP)* (2018), pp. 1–5, http://dx.doi.org/10.1109/ICCCSP.2018.8452820
37. V. Vippalapalli, S. Ananthula, Internet of things (iot) based smart health care system, in *2016 International Conference on Signal Processing, Communication, Power and Embedded System (SCOPES)* (2016), pp. 1229–1233, http://dx.doi.org/10.1109/SCOPES.2016.7955637
38. K. Natarajan, B. Prasath, P. Kokila, Smart health care system using internet of things. J. Netw. Commun. Emer. Technolog. (JNCET) **6**(3) (2016), https://pdfs.semanticscholar.org/34bc/7eeeeda54fc47c6467d01196e5d02df21bbb.pdf
39. N. Pulur, D. Altop, A. Levi, A role and activity based access control for secure healthcare systems, in *Information Sciences and Systems 2015*, ed. by O.H. Abdelrahman, E. Gelenbe, G. Gorbil, R. Lent, Lecture Notes in Electrical Engineering, vol. 363 (Springer International Publishing, Berlin, 2016), pp. 93–103, http://dx.doi.org/10.1007/978-3-319-22635-4_8
40. A. Ranjan, G. Somani, Access control and authentication in the internet of things environment, in *Connectivity Frameworks for Smart Devices*, ed. by Z. Mahmood, Computer Communications and Networks (Springer International Publishing, Berlin, 2016), pp. 283–305, http://dx.doi.org/10.1007/978-3-319-33124-9_12
41. S. Alshehri, R.K. Raj, Secure access control for health information sharing systems, in *2013 IEEE International Conference on Healthcare Informatics* (2013), pp. 277–286. https://doi.org/10.1109/ICHI.2013.40
42. M. Haque, A.-S.K. Pathan, C.S. Hong, Securing U-healthcare sensor networks using public key based scheme, in *2008 10th International Conference on Advanced Communication Technology* (IEEE, 2008), pp. 1108–1111, http://dx.doi.org/10.1109/icact.2008.4493960
43. C.-T. Li, T.-Y. Wu, C.-L. Chen, C.-C. Lee, C.-M. Chen, An efficient user authentication and user anonymity scheme with provably security for IoT-based medical care system. Sensors **17**(7), 1482+ (2017), http://dx.doi.org/10.3390/s17071482
44. V. Adat, B.B. Gupta, Security in internet of things: issues, challenges, taxonomy, and architecture. Telecommun. Syst. **67**(3), 423–441 (2018). https://doi.org/10.1007/s11235-017-0345-9
45. A. Sehgal, V. Perelman, S. Kuryla, J. Schonwalder, Management of resource constrained devices in the internet of things. IEEE Commun. Mag. **50**(12), 144–149 (2012), http://dx.doi.org/10.1109/MCOM.2012.6384464
46. C.N. et al., The kerberos network authentication service (v5), ietf rfc 4120 (2005), https://tools.ietf.org/html/rfc4120. Accessed 20 April 2017

47. H. Kim, E.A. Lee, Authentication and authorization for the internet of things. IT Professional **19**(5), 27–33 (2017), http://dx.doi.org/10.1109/mitp.2017.3680960
48. P.P. Pereira, J. Eliasson, J. Delsing, An authentication and access control framework for CoAP-based Internet of Things, in *IECON 2014 - 40th Annual Conference of the IEEE Industrial Electronics Society* (IEEE, 2014), pp. 5293–5299, http://dx.doi.org/10.1109/iecon.2014.7049308
49. Google beacons, Technical Report, https://developers.google.com/beacons/
50. Axiomatics, Attribute based access control (abac), https://www.axiomatics.com/attribute-based-access-control/. Accessed 10 Sep 2017
51. E. Coyne, T.R. Weil, Abac and rbac: scalable, flexible, and auditable access management. IT Professional **15**(3), 14–16 (2013). https://doi.org/10.1109/MITP.2013.37
52. Z. Mao, N. Li, W. Winsborough, Distributed credential chain discovery in trust management with parameterized roles and constraints (short paper), in *Information and Communications Security*, ed. by P. Ning, S. Qing, N. Li, Lecture Notes in Computer Science, vol. 4307 (Springer, Berlin, 2006), pp. 159–173, http://dx.doi.org/10.1007/11935308_12
53. D. Schwartmann, An attributable role-based access control for healthcare, in *Computational Science - ICCS 2004*, ed. by M. Bubak, G. van Albada, P. Sloot, J. Dongarra, Lecture Notes in Computer Science, vol. 3039 (Springer, Berlin, 2004), pp. 1148–1155, http://dx.doi.org/10.1007/978-3-540-25944-2_149
54. extensible access control markup language (xacml) version 3.0 (2013), http://docs.oasis-open.org/xacml/3.0/xacml-3.0-core-spec-os-en.html. Accessed 02 Nov 2017
55. S. Chandra, S. Paira, S.S. Alam, G. Sanyal, A comparative survey of symmetric and asymmetric key cryptography, in *2014 International Conference on Electronics, Communication and Computational Engineering (ICECCE)* (2014), pp. 83–93

Chapter 4
System Implementation and Evaluation

Abstract This chapter presents a detailed description of the implementation of the proposed access control architecture that is presented in Chap. 3. This chapter aims to show how the proposed architecture can achieve significant improvement by reducing the number of policies required for specifying access control settings while providing efficient access control in an IoT environment. Standard XACML [1] based policy scheme is used. The proof of concept prototype is implemented using a physical testbed experiment to demonstrate the feasibility of the proposed approach. Note, the system's operations are tested with both symmetric and asymmetric key cryptography based approaches.

4.1 Introduction

The aim of the system implementation is to demonstrate the feasibility of the proposed access control architecture using a physical experimental platform. In this section, first, a testbed development is discussed and then the employed methodology for evaluation is explained. We employ the same notations that are used to discuss various architectural components in Chap. 3.

4.1.1 Testbed Development

To conduct the proposed experiment in a physical testbed, it is necessary to develop three major components of the system, namely, the resourceful server (i.e. the central system), the client (i.e. the user devices) and the resource-constrained server (i.e. IoT devices). Figure 4.1 illustrates the testbed set up. The CMS (i.e. the resourceful server) is implemented on a MacBook Pro powered with a dual-core 2.4 GHz Intel processor and 4 GB of DDR3 RAM running the latest High Sierra MacOS. All components of the CMS are developed in C# using Microsoft's .NET development framework. The authentication and authorization platform runs as services in the background

S. Pal, *Internet of Things and Access Control*, Smart Sensors, Measurement and Instrumentation 37, https://doi.org/10.1007/978-3-030-64998-2_4

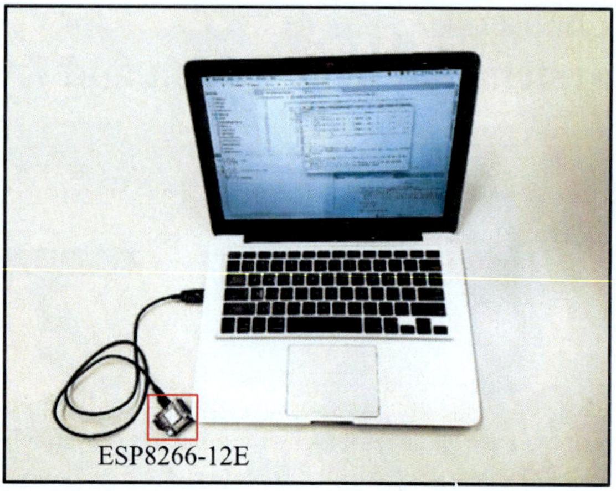

Fig. 4.1 A sample psychical testbed used for the experiments

and replies to requests from multiple sources including direct TCP connection or MQTT [2].

Significant to note that, the implementation could be achieved using various communication protocols e.g. HTTP, MQTT, CoAP, bare TCP, etc. [3, 4], however, in this experiment we have mainly used MQTT. HTTP is document-centric while MQTT is data-centric. MQTT is chosen for the communication with the THs as it is light-weight and easily scalable using its publish/subscribe messaging protocol suitable for resource-constrained devices. The MQTT protocol also ensures high delivery guarantees with its three levels QoS. In contrast, HTTP uses lengthy headers (using text message format) that are unnecessary for our use case. MQTT includes a short message header as the smallest packet message size is 2 bytes [5]. CoAP is widely used for IoT protocols, but MQTT is a many-to-many communication protocol where CoAP is a one-to-one protocol. While CoAP has a lower protocol overhead, MQTT's design is more attractive for large deployment as it readily improves scalability and robustness (redundancy can be achieved by simply using multiple MQTT brokers in bridge mode). A thorough comparison of CoAP versus MQTT can be found in [6].

For implementing the PDP, an open-source project called *abc.xacml* [7] is used. It provides a ready to use .Net library. The implementation for the PEP is customized by writing the policies as per the requirements specific to the use-case. To achieve a generic and light-weight implementation, MySQL databases are used to store persistent data (e.g. policies, user and resource information, attributes, and similar). In contrast to the other capability designs presented in the literature (e.g. [8]), which typically follow a heavy-weight XML structure, JSON is employed to represent and store the capabilities.

For the purpose of the experiments the clients representing UDs are installed directly on the same hardware as the CMS to reduce unnecessary communication through the network. Note that, the clients can also be run on mobile devices e.g. smartphones or tablets. For the resource TH (i.e. the resource-constrained device), an ESP8266-12E microcontroller is used (cf. Fig. 4.2). These are low cost devices

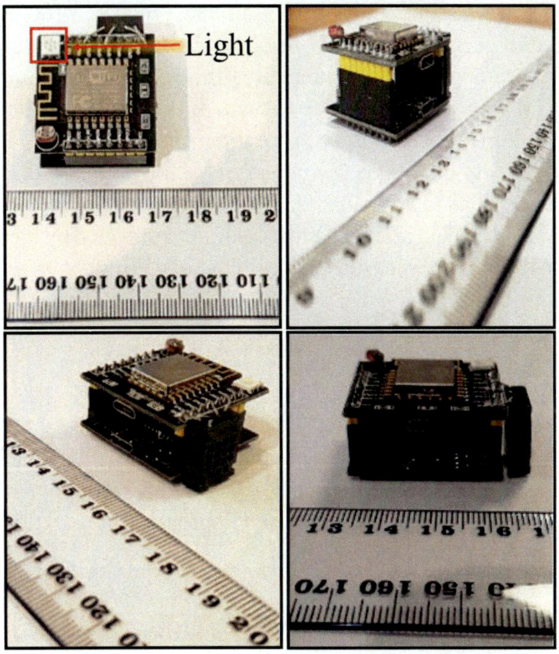

Fig. 4.2 ESP8266-12E used for the experiments

that allow mass production of connected devices for general consumer use (one microcontroller costs around 2.5 Australian Dollars). They are highly optimized to guarantee a moderate level of performance with low power and low memory consumption. The ESP8266-12E has a ready to use WiFi connection and fully supports the TCP/IP stack. It consists of a 32-bit RISC CPU with scalable speed, 60–160 MHz, 42 KB of RAM and 4 MB of flash memory. Each TH is running a domain specific software on top of NodeMCU [9], a 'Lua'-based interpreter for the ESP8266-12E. Lua is a high-level language that greatly facilitates design and implementation but the code it produces will not be optimized, so the results obtained are conservative. AES-CBC (128 bits key size) is employed for secure communication.

4.1.2 Methodology

In this experiment, the tests measure both the time from when a UD requests a capability from the CMS (after detecting a broadcast signal from a TH) to when the TH responds to the access request after the UD provides the capability and important segments of that process. The experimental setup uses a database that contains 1000 random users, each associated with specific permissions, and 5000 THs registered in the TRR. In order to demonstrate the feasibility of the model, and to ensure reliable data, each test was run 100 times. All tests involving the

TH were sequential (i.e. the same test repeated 100 times where request parameters have been randomized) because the networking capabilities of the ESP8266-12E micro-controller is moderately limited. The communication between the UD and the CMS are mostly concurrent which demonstrates the ability of the CMS to handle multiple requests at the same time. On each test, the UD randomly selects the TH to access to guarantee no impact of caching, etc., within the databases. This allows us to demonstrate the load/delay in our implementation. The UD sends access requests that turned on and off a physical light controlled by the TH that can be seen in Fig. 4.2. All times are measured in milli-seconds (ms).

4.2 Results

This section details the obtained results for the experiment based on the following two approaches, first, symmetric key based approach (discussed in Sect. 4.2.1) and second, asymmetric key based approach (discussed in Sect. 4.2.2).

4.2.1 Performance Analysis: Symmetric Key Approach

Now a detailed performance analysis of the system is presented followed by the discussion described in Sect. 3.6 of Chap. 3. The results include the following phases, they are: (1) full communication from a capability request to a resource access, (2) performance of the CMS, and (3) processing of an access request by the TH. These steps are depicted in Fig. 3.14 of Chap. 3. The compiled Lua program installed on the TH occupies 18 KB of ROM and uses about 26 KB of RAM. The amount of RAM used depends on the state of the TH and more RAM is used while processing requests or publishing messages via MQTT. Recall, that the ESP8266-12E has 42 KB of RAM in total. Importantly, for this experiment, a valid capability is used.

In Fig. 4.3, we depict the overall processing time (the black curve, labelled as 'Duration') of the system from when a UD requests a capability to when it obtains access to a TH (we assume that in this case the request is evaluated as a valid request). The average time taken is 1010 ms (with a standard deviation, denoted as σ, of 200 ms). Returning to Fig. 3.14 (cf. Chap. 3), steps 2–7 are executed in this case. One hundred access requests are sent from UD to turn on and off a physical light controlled by the TH.

Interestingly, we observed that there are large fluctuations in the time taken. We examined these fluctuations and found that they are neither from the ESP8266-12E nor network latency as we initially suspected. Instead, they are due to the database access required during each capability request evaluation to collect attributes. We use a standard installation of MySQL without any specific optimization. The red curve in Fig. 4.3, labelled by 'Decryption', shows the amount of time each TH spends on decryptions alone. On an average, for a single access request, each TH spends

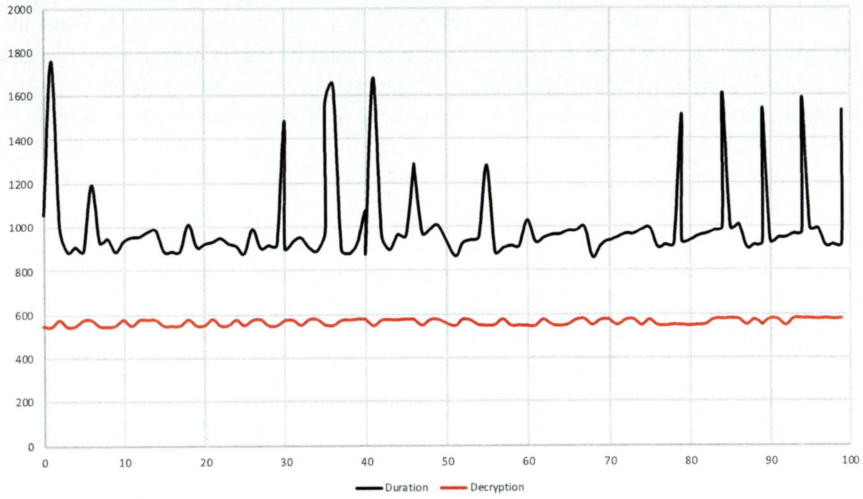

Fig. 4.3 Processing of 100 access requests each with a fresh capability. The X axis represents the message_id and the Y axis represents the time (ms)

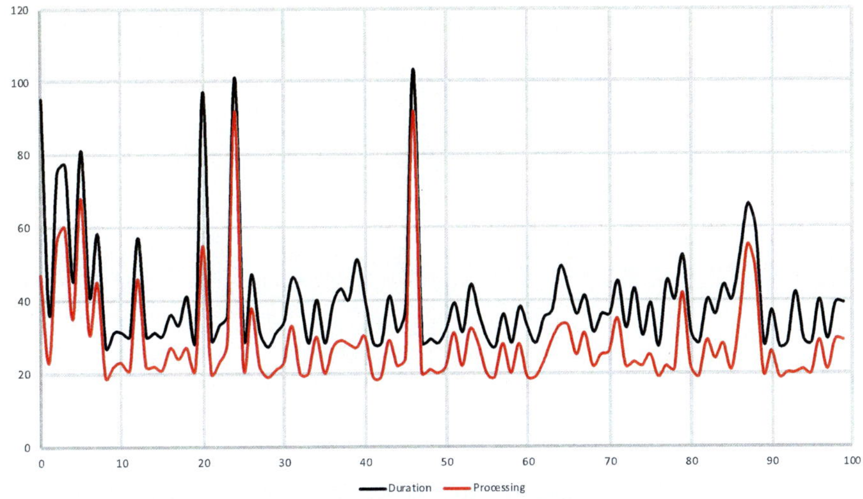

Fig. 4.4 Processing of 100 capability requests by the CMS. The X axis represents the message_id and the Y axis represents the time (ms)

561 ms ($\sigma = 13$ ms) decrypting the capability and access tokens. The shape of the curve illustrates that the measurement is very stable. It can be seen that this operation is the largest single contributor to the overall time.

In Fig. 4.4, we show the time taken for each of the 100 capability requests to be processed by the CMS ('Duration' in black curve). Note that this is a portion of the time shown in Fig. 4.3. The average time taken in this case is 40 ms ($\sigma = 16$ ms).

This includes receiving a capability request (i.e. an incoming message) from the UD, performing a lookup on the database for attributes, generating and evaluating an XACML request from these attributes, generating a capability, composing and encrypting the message and finally, sending the capability within the encrypted reply message to the UD. Returning to Fig. 3.14 (cf. Chap. 3), steps 2–5 are evaluated in this case. The 'Processing' in the red-curve of Fig. 4.4 illustrates the actual processing of an XACML request starting from the collection of the attributes to the evaluation within the PDP engine. The processing can last from 18 to 92 ms. However, we noticed that, once all required attributes have been collected it then takes an average 7 ms for processing concurrent XACML requests. In practice, this will also reduce the overall processing time of the CMS.

Note, there is a strong correlation between the processing time and the total duration of (the round-trip) communication for a capability request. The most expensive task during the processing is the generation of the XACML requests where attributes need to be collected from multiple sources, which is a MySQL server in our case. Figure 4.3 furthermore demonstrates the fact that this can substantially affect the overall processing time.

In Fig. 4.5, we show the cumulative number of processed requests. The figure illustrates that the requests are processed almost as soon as they arrive at the CMS because the curves 'Req CMS' and 'Req UD' are almost indistinguishable. The

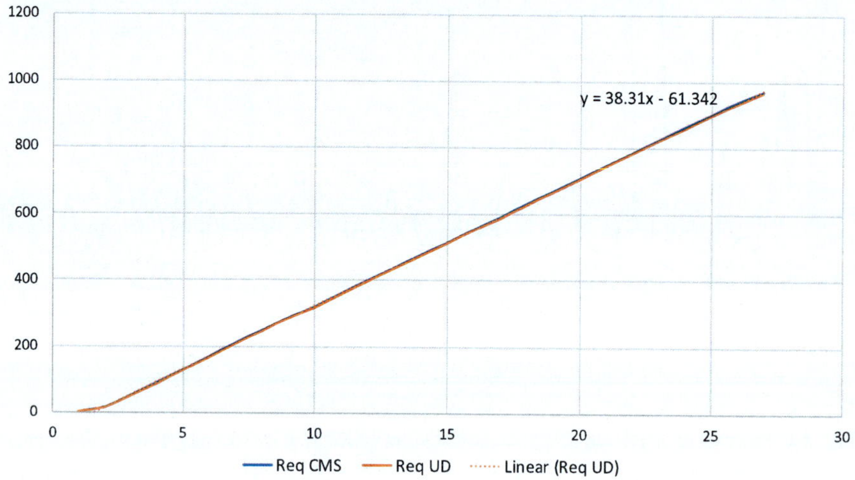

Fig. 4.5 Requests processed by the CMS over time. The X axis represents time in seconds and the Y axis represents the cumulative number of requests received and processed. "Req CMS" (resp. "Req UD") is the cumulative number of requests (resp. reply) received by the CMS (resp. UD). "Linear (Req UD)" is a linear fit over "Req UD"

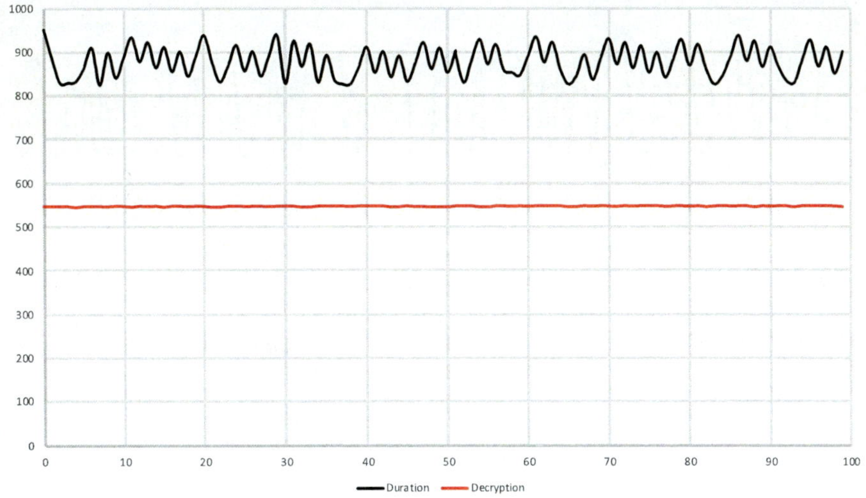

Fig. 4.6 Processing of 100 access requests by the TH with a cached capability held by the UD. It starts from when a UD sends a request to the TH and receives a reply back from the TH. The X axis represents the message_id and the Y axis represents the time (ms)

linear trend-line shows that, given this particular hardware, the server is able to fully process around 38 requests per second.[1]

In Fig. 4.6, we illustrate the total duration from when a UD sends a request to the TH and receives a reply from the TH ('Duration' in black-curve) for the situation where the UD already holds an appropriate capability. Returning to Fig. 3.14 (cf. Chap. 3), steps 6 and 7 are evaluated in this case. The average time taken in this case is 877 ms ($\sigma = 34$ ms). This time-span covers the following three phases: (1) network communication latency between the UD and the TH, (2) encryption and decryption of messages by both the UD and the TH, and (3) processing of a decrypted request by the TH. A major portion of the overall 877 ms is however used on cryptographic computations by the TH. Note, in this case, we start with a cached capability within the UD and reuse this for each and every access request.

Once again, for an overall time comparison, in Fig. 4.6, we show the decryption time taken by the TH ('Decryption' in red-line). Note that the difference between the decryption times shown in Figs. 4.3 and 4.6 are due to the use of a cached capability for Fig. 4.6 whereas a new capability is generated for each request in Fig. 4.3. Hence, there is a lack of fluctuation seen in Fig. 4.6 compared to Fig. 4.3. However, it can be seen that the operations are taking very similar time. Again, it can be noticed that the decryption is the most time-consuming operation.

[1]Note, this is the performance on a consumer grade computer and we can get much better performance running the exact same setup on more powerful hardware. Also, we have implemented a synchronous TCP server that handles requests submitted to the CMS. An asynchronous implementation would also boost the performance considerably. These improvements are however out of the scope of our contribution in this chapter.

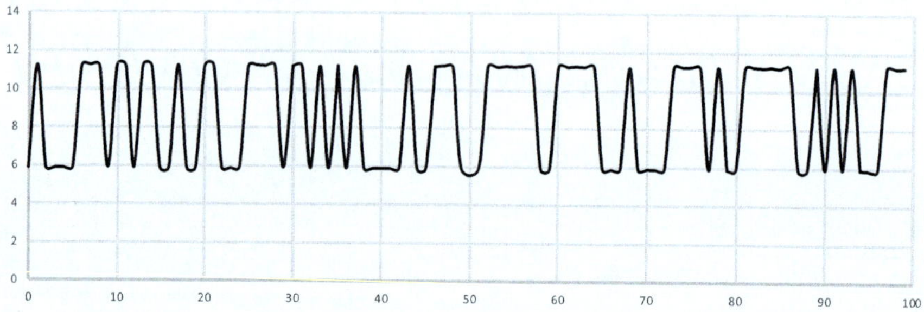

Fig. 4.7 Processing of 100 decrypted access requests by the TH (i.e. parts of the Algorithm 3.1 of Chap. 3). The X axis represents the message_id and the Y axis represents the time (ms)

The network capability of the ESP8266-12E is largely responsible for the fluctuation observed in the black curve of Fig. 4.6, although the variation is not as acute as seen in Fig. 4.3. Recall, we use the MQTT protocol for communication between the UD and the TH. Still, the system is spending an average 102 ms simply processing TCP/IP packets and most of this time is spent on the TH. Note, encryptions and decryptions are also performed by the UD but this device (i.e. the UD) is reasonably powerful and therefore these operations only make up a small fraction of the overall time shown in the figures.

In Fig. 4.7, we illustrate the processing time for different parts of Algorithm 3.1 (discussed in Chap. 3) excluding the decryption and encryption stages (i.e. execution of only steps 4–8 of Algorithm 3.1). It takes an average 9 ms for the TH ($\sigma = 2.5$ ms). In this case, a request is evaluated as successful and the state of the light attached to the ESP8266-12E is changed depending upon the access request.

We observed that, if the state of the light does not need to change, then it takes an average 6 ms to process an access request. If the state of the light needs to be changed, then it takes an average 11.5 ms to process the request because the TH must physically change state. We argue that this extra delay is particularly important for physical sensors, e.g. temperature or pressure sensors, because it usually takes more than a few tens or hundreds of milliseconds to accurately measure these quantities. In our experiment, the requested state for the TH is chosen randomly.

4.2.2 Performance Analysis: Asymmetric Key Approach

We argue that light-weight cryptosystems are likely most suitable for resource constrained IoT devices. Therefore, in the previous section (Sect. 4.2.1), we discussed the employment of a light-weight protocol that relies on symmetric key cryptography and provided a detailed discussion of suitability of using a symmetric key based approach for our proposed access control architecture. However, in Chap. 3, we also noted that our architecture can be implemented using either symmetric or asymmetric key cryptosystems. To show that the proposed architecture could easily work with

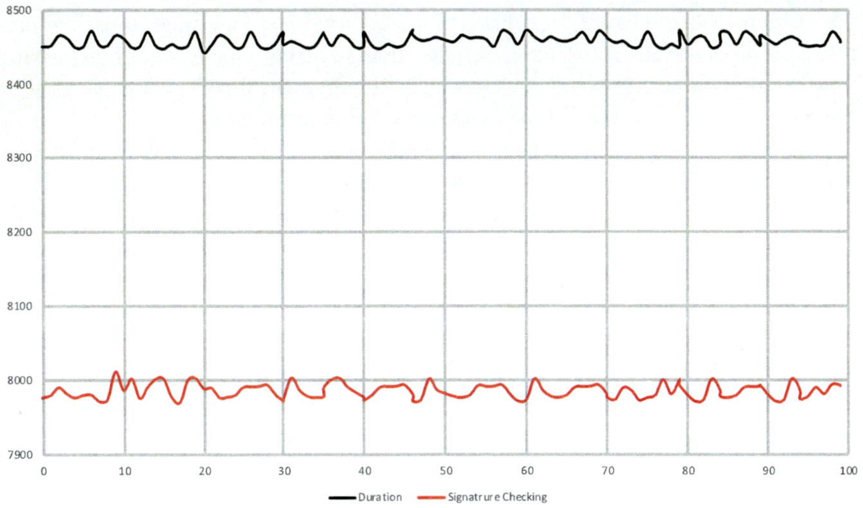

Fig. 4.8 Processing of 100 access requests each with a fresh capability. The X axis represents the message_id and the Y axis represents the time (ms)

either approach, in this section, we used public key cryptography for guaranteeing the authenticity of information between different components within the architecture. Recall, implementing public key cryptosystems on IoT devices can be challenging due to the resource-constrained nature of the devices. But it can provide adequate security where resource limitation is not an issue [10].

Now we present a detailed performance analysis of the system that we designed in Sect. 3.7 of Chap. 3. In Fig. 4.8, we show the overall processing time (the black curve, labelled as 'Duration') of the system from when a UD requests a capability to when it obtains access to a TH with a valid request. We noted that the average time taken is 8457 ms ($\sigma = 8$ ms). Returning to Fig. 3.15 (cf. Chap. 3), steps 2–7 are executed in this case. One hundred access requests are sent from UD to access the TH. The red curve in Fig. 4.8, labeled by 'Signature Checking', illustrates the amount of time each TH spends on the signature checking alone. Note, in this case, two signatures one from the CMS and another from the UD are checked. On an average, for a single access request, each TH spends 7984 ms ($\sigma = 9$ ms) for signature checking. The difference (in time) between the black curve (labelled as 'Duration') and the red curve (labelled as 'Signature Checking') is all the other required operations e.g. the CMS creating and signing the new capability and the TH checking the decrypted capability. These operations take approximately 460 ms.

Note that this result is obtained using an RSA key size of 64-bit. We observed that in Fig. 4.8 this higher amount of time taken due to the resource constrained nature of the ESP8266-12E. However, we argue that it is not a practical and secure implementation (we provide more discussion in Sect. 4.3). We only try to demonstrate the practicality of the implementation in a physical testbed. In this particular experiment (i.e. asymmetric key based implementation), we are most interested to see the performance of constrained IoT devices rather than focusing on the overall processing time (i.e. the round-trip time) of the system. Importantly, this design can easily be implemented where resource limitation is not a decisive factor.

Therefore, we present the results of the performance evaluation of the most relevant parts of our system—the communication between the UD and the TH and the evaluation of a capability by a TH. As the THs are the most resource-constrained elements of the system it is important to ensure that the requirements of our architecture do not pose an unmanageable load upon them. That said, we exclude signature checking for the following evaluation results.

We examine a number of scenarios, from those described in Sect. 3.5 of Chap. 3, including both when access is granted and when it is refused and involving the TH checking a varying number of conditions. Recall, in order to demonstrate the feasibility of our model, each test was run 100 times to ensure reliable data. To demonstrate the time taken by our proposal, all considerations which are extraneous are excluded, e.g. delays due to other network traffic. This allows us to demonstrate the load/delay created by our proposal. For every success and fail, an appropriate message (e.g. granted or refused) is sent to the UD. Once again, note that all times are shown in milli-seconds (ms).

Figure 4.9 shows the case where the capability fails on the initial test of Algorithm 3.2 (discussed in Chap. 3), whether the current time falls within the valid period defined by the issued and expiry time fields of the capability. This represents the minimum time for a response by a TH once a UD presents it with a capability. The results show that the median time taken for this stage is 0.542 ms ($\sigma = 0.003$).

Figure 4.10 shows two results (i) when the time stamp is valid but the capability does not apply to the TH. The median time taken for this stage is 0.564 ms ($\sigma =$

Fig. 4.9 The capability authorization time for an invalid capability. In this case, the time stamp (ts) is not valid, therefore, the capability is rejected at the very first instance without checking other fields

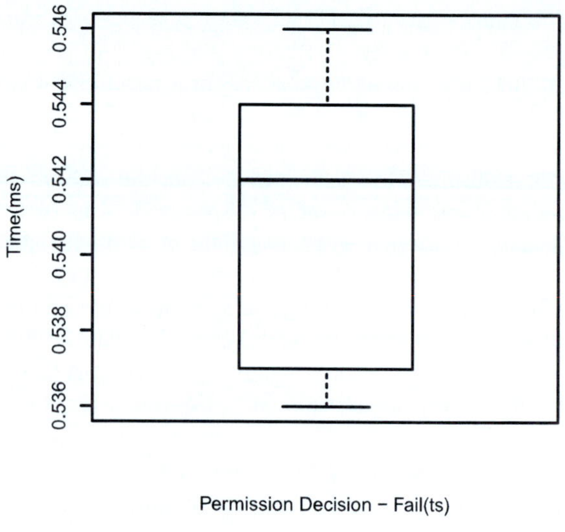

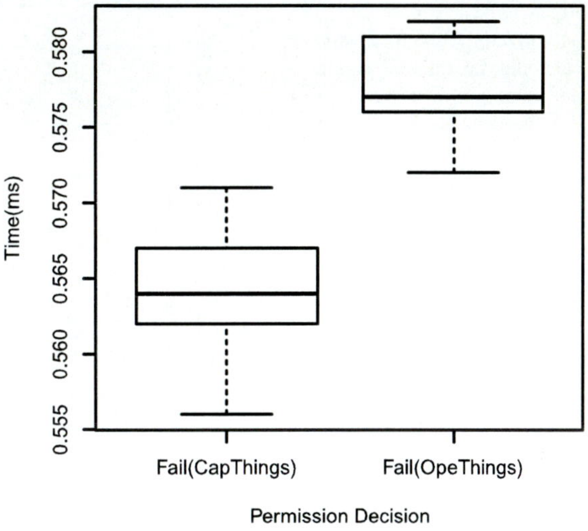

Fig. 4.10 The capability authorization time for an invalid capability. 'Fail(CapThings)' denotes that the capability does not apply to the TH. 'Fail(OpeThings)' denotes that the capability does not allow a valid operation on the TH. Note, in both the cases the time stamp is valid, i.e. the capability is valid for a certain period of time

0.003). And (ii) when the time stamp is valid, the capability applies to the TH but the capability does not allow the requested operation on the TH. The median time taken for this stage is 0.577 ms ($\sigma = 0.003$). These results are only slightly longer than those in Fig. 4.9. The second result is longer than the first as the test for the capability applying to TH must be carried out and passed before the test on whether the requested operation is allowed by the capability is applied.

In Fig. 4.11, we illustrate the results of checking a single condition. The right-hand result is when the condition is failed, the left hand result is when the condition is

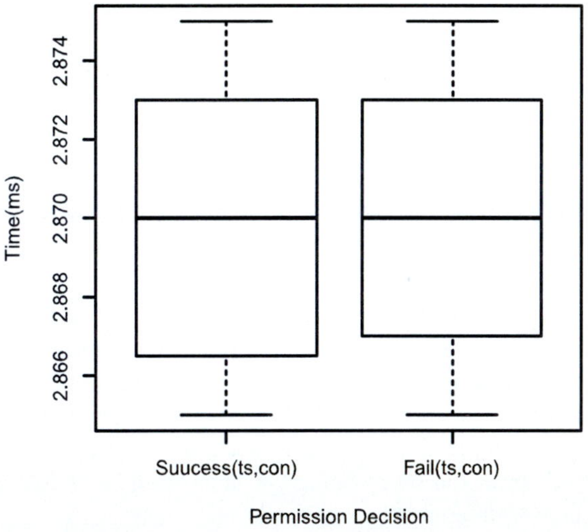

Fig. 4.11 The capability authorization time for condition success and failure. 'Success(ts,con)' represents the time taken when condition returns true after all previous checks have succeeded. Similarly 'Fail(ts,con)' represents the time taken when the condition is not valid but all previous checks succeeded

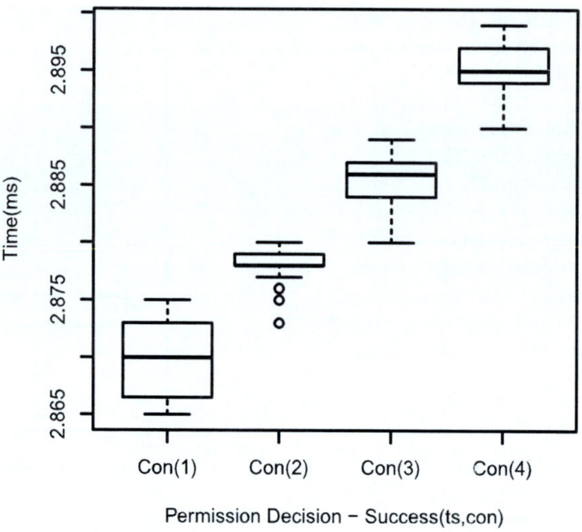

Fig. 4.12 The capability authorization time for a valid capability i.e. the time stamp and condition(s) are valid. In this case we vary the number of conditions from one to four

passed. For comparison purposes signature checking was excluded. All comparable approaches use signature checking, typically the same number of checks. Signature checking is time consuming and therefore heavily affected by device power. Recall, to focus on the more relevant parts of our design, we exclude signature checking at this point. We highlight that this time here is considerably longer than the previous cases, although still only a handful of milli-seconds, as there is some setup involved in condition checking. The condition checked here was whether the location of the TH, represented as a string stored in the TH, satisfied the requirement expressed in the '*Condition Rule*' that we discussed in Sect. 3.4.6 of Chap. 3. The median time required when the condition evaluates to true is 2.87 ms ($\sigma = 0.003$) and when the condition evaluates to false is 2.87 ms ($\sigma = 0.003$). That the results are the same (at least to the precision shown here) is to be expected, as all other fields must be checked before this test and whether a condition succeeds or fails will require much the same operations. The median RAM and ROM required for the both cases are 15 KB and 9.2 KB respectively.

Figure 4.12 shows the results when the number of conditions to be checked are varied between one and four. In other words, we use different number of conditions in the capability to see the overall performance of the TH. For the simplicity to the experiment, in all cases, all conditions returned true to enable checking to complete. Note that the first condition is the same as in Fig. 4.11 for comparison purposes. The second condition checked whether the current time was within a period specified in the '*Condition Rule*' (as discussed in Sect. 3.4.6 of Chap. 3), the third condition checked the date and the fourth condition involved checking the remaining battery power in the TH. Note that checking of extra conditions after the first added very little time, the set up being the same in all cases. We have observed that the median time required for the one condition checking is 2.87 ms ($\sigma = 0.003$), two conditions is 2.87 ms ($\sigma = 0.003$), three conditions is 2.88 ms ($\sigma = 0.003$) and for the four conditions is 2.89 ms ($\sigma = 0.003$). Recall, as we discussed above, we have excluded signature checking. We have noted that, the total time for a response to the UD is less than three milli-seconds.

4.3 Comparison with Existing Approaches

In the above experiments, we used both symmetric key cryptography and asymmetric key cryptosystems. Partly this was done as no comparable results for a symmetric key approach exist in the literature. Indeed, even where timings are given for other approaches comparisons are difficult, due to the vastly different power of the THs used in the experiments. This will affect the speed at which operations are performed, and is particularly noticeable in the cryptographic operations, which are typically the most time consuming steps.

The situation was further complicated by the paucity of asymmetric key cryptographic implementations for the ESP8266-12E. Given its resource constrained nature the only implementation available employs a 64-bit key size. This does not support practical and secure implementations. Therefore, to gain some comparison, we tested our data structures with both RSA and AES on a relatively high speed platform (a MacBook Pro, dual core 2.4 GHz Intel processor and 4 GB of DDR3 RAM running the latest High Sierra MacOS). On that platform, we tested the cryptographic operations using the same messages and data structures as in our symmetric key implementation. We ran 1000 samples, using an RSA key size of 2048-bit, AES key size of 128-bit and SHA256 Hashing.

We detail our results in Table 4.1. Note the difference in the two rightmost columns in the table (i.e. AES(SA) and AES(FA)). This difference is explained by the caching of the operations. Columns 3 and 4 (i.e. RSA(Sg) and RSA(Ve)) are included for comparison purposes only, as they represent the time taken by operations that would occur on the CMS (composition and signing of the encrypted messages). The most important comparisons are then between the totals of columns 1 and 4 (i.e. RSA(En) and RSA(Ve)), the recovery of the signed hash and the hashing of the received message respectively and, conservatively, column 6. This gives a ratio of between 2

Table 4.1 Comparison of RSA and AES implementation. Where, msgL = message length, RSA(En) = RSA(Encrypt), RSA(De) = RSA(Decrypt), RSA(Sg) = RSA(Sign), RSA(Ve) = RSA(Verify), AES(SA) = AES(Successive Access), AES(FA) = AES(First Access). All times are measured in ms

msgL	RSA(En)	RSA(De)	RSA(Sg)	RSA(Ve)	AES(SA)	AES(FA)
128	1.691	29.467	28.693	2.084	0.0061	1.701
245	3.179	58.173	27.927	2.024	0.0073	–
256	3.299	55.875	27.971	2.018	0.0085	–
1024	11.062	134.74	27.971	2.051	0.0243	2.001

and 6 for the asymmetric to symmetric approaches. Given the time shown in Figs. 4.3 and 4.6 for cryptographic operations on the TH, we can see that the time saving on constrained devices for the symmetric approach can be considerable.

We acknowledge that this time saving decreases as the power of the TH increases. While there are a number of other proposals for IoT access control based on CapBAC, only one, [11], to our knowledge, gives detailed timing results for an implementation. That proposal is based on asymmetric key cryptography. In [11], the statement is made *"the time required for the whole mechanism was 1205.83 ms"* of which approximately half is taken up by signature checks on the TH (i.e. approximately 600 ms). Note that these results were obtained on a more powerful TH than the one used for our experiment. If our approach was implemented on the hardware used in [11], we would expect a significant time saving. More importantly, note that the 'non-signature' checking portion of our implementation takes approximately 460 ms (cf. Fig. 4.8) while that of [11] takes approximately 600 ms, a significant saving considering that our TH hardware is less powerful.

Proposal [11] involves two checks, one for the UD's signature on the request and another for the CMS (or equivalent) signature on the capability. The best times in the literature for these signature checks uses ECC. Including the time period for two such checks in our results would mean the signature checks, which are not the focus of our asymmetric key based implementation, would completely overshadow the time taken by the functions of our proposal. This does show that our proposal using an asymmetric key cryptography adds no significant time to the basic signature checks required by all CapBAC proposals for the IoT.

4.4 Comparison of Number of Policy Expressions

In this section, we provide a comparison of the number of policy expressions required in our proposed approach in a real-world scenario with the existing approaches. To illustrate the reduction in the number of policies flowing from our approach, we consider a use case example of IoT-enabled smart healthcare facility where serval actors are involved (cf. Sect. 3.3 of Chap. 3). In such a scenario there are many patients, doctors, nurses and corresponding medical devices are attached to the patients. We assume that on average there are 10–15 devices connected per bed [12]. These devices can be of many types e.g. blood pressure sensors, blood analyzers, body pressure sensors, etc. [13]. A typical hospital will employ one nurse for every four beds and one doctor for every three nurses (or twelve beds) [14]. If we consider a facility with 1200 beds, there will be approximately 300 hundred nurses and 100 doctors [15].

In a typical hospital, nurses and doctors are assigned to wards. Let us assume, for the sake of argument that the ward size is twenty beds, nurses are assigned to individual wards and doctors cover two wards or equivalent.

Access could simply be granted to all staff to all sensors of all patients and currently this approach is often adopted for lack of an alternative. However, this is not fine-grained and risks patient privacy and confidentiality. Addressing these issues require a fine-grained approach to ensure that staff only have access to the data of patients under their care. In the absence of any mechanism for managing policies, the policies would have to be written on a one-to-one basis. That is, each staff member would have to be given access to the sensors on an individual basis. Given the above figures, the number of policies would be given by:

(a) 20 (number of patients per ward) × 10 (number of sensors per patient) × 300 (number of nurses) = 60,000.

(b) 20 (number of patients per ward) × 10 (number of sensors per patient) × 2 (number of wards allocated to each doctor) × 100 (number of doctors) = 40,000.

That is, fine-grained access control in the scenario described above would require 100,000 individual policy expressions in the absence of any policy management. Not only would this be almost impossible to compose, it would be very hard to manage when patients or staff are moved from ward to ward or when the sensors attached to a patient are altered. Note that, the above is a conservative estimate. The number of sensors per patient could soon reach as high as 65 [16].

Now consider the alternative under our proposal. Each sensor is registered in the system, with an attribute identifying the patient to whom it is attached. Doctors, nurses and patients also have attributes identifying their status and ward.

Each job function within the organization (e.g. registrar, different types of specialist e.g. cardiologist, neurologist, anaesthetist, nurse, senior nurse. etc.) will have a corresponding *role*. The number of such roles may be in the order of a dozen or so. Each role will need a '*Role Membership Rule*', based on the attributes of the potential member. As noted above, the devices attached to patients will come in a number of different forms. Each role can be given a rule covering each type, to allow tailoring to the available operations of the different types. Policy management can be achieved by employing capability templates and parameterization as described in Sect. 3.4.7 of Chap. 3. Capability templates specify the type of device and the operations conveyed by role membership. Recall, capabilities are issued from the appropriate capability templates. The operations will vary from role to role, but this does not affect the number of overall policies, so is ignored here. For example, consider a rule governing 'heart sensors'. We would then have a capability template of the form e.g.,

(*Type*: Heart Sensors; *Operation*: Read; *Parameterization Rule*: If Device.Patient.Ward = User.Ward)

That is, a user that has fulfilled the role membership requirements will be granted a capability for the sensor attached to a patient if the role member's ward matches the ward of the patient to whom the sensor is attached.

One such rule could be provided for each type of sensor. As noted, this will allow the available operations to be tailored to the precise device type and role of the user. The number of policies required is then:

(a) One '*Role Membership Rule*' for every role.

(b) One '*Parameterization Rule*' for every sensor type for every role.

Assuming, as above, twelve roles and, for example, ten sensor types, we have a total of 132 policy expressions. This is significantly less than the alternative. Note also that updates are significantly easier as patients and staff assignment to wards change; all that is required is changes to the attribute assignment. Access will then flow from there. In the alternative case, individual alteration to numerous policies will be required.

We note that the above is an example only, as there other methods of structuring the attribute assignment (for example, doctor assignment to patients rather than wards) however, the savings would also be significant in that case.

4.5 Adversary Analysis

Any IoT system presents a large attack surface to potential adversaries [17, 18]. As we discussed earlier (in Chap. 2) possible vulnerabilities extend from the devices themselves, through the communication between them to the services and applications provided. Users, and the inherent characteristics of IoT systems, also provide threat opportunities. However, a detailed analysis on IoT threats and attack modelling is out of the scope of this chapter. In this section, we follow the approach of [19] and outline a few adversary scenarios and their countermeasures that overlap with our use-case as outlined in Sect. 3.3 of Chap. 3.
Example Scenarios: The possible ways for an adversary to misuse the system include impersonating the end user, unauthorized access to THs, unauthorized use of attributes, etc. Now we discuss that an attacker or an authorized user could attempt the following:

- *Scenario 1:* An adversary attempts to use attributes that have not been assigned to them in an attempt to access a resource.
- *Scenario 2:* An adversary attempts to gain access to a resource in violation of the capability conditions.
- *Scenario 3:* An adversary attempts to use a capability that is not assigned to them in an attempt to access a resource.
- *Scenario 4:* An adversary attempts to modify a capability.
- *Scenario 5:* An adversary attempts to gain access to a TH that is not authorized by the capability.

Addressing the Scenarios: We argue that asymmetric key cryptography can provide better security compared to a symmetric key cryptography. However, implementing public-key cryptosystems on IoT devices is challenging due to the resource-constrained nature of the devices. Therefore, in this adversary analysis, we examine the use of a light-weight protocol that relies on symmetric key cryptography (cf. Sect. 3.6 of Chap. 3). Recall, our proposed access control architecture could easily work with either approach. We further assert that, a system that is developed based on a symmetric key cryptography can easily be built using an asymmetric key cryptosystem.

```
<Policy xmlns="" RuleCombiningAlgId="rule-combining-algorithm:deny-overrides"
PolicyId="policy1">
<Rule Effect="Permit" RuleId="read-resource:rule1">
  <Description>
    Doctors can only read their patient's heart sensor.
  </Description>
  <Target>
    <AnyOf>
      <AllOf>
        <Match MatchId="function:string-equal">
          <AttributeValue DataType="string"> get </AttributeValue>
          <AttributeDesignator AttributeId="action:action-name" Category="action"
          DataType="string"/>
        </Match>
        <Match MatchId="function:string-equal">
          <AttributeValue DataType="string"> heart sensor </AttributeValue>
          <AttributeDesignator AttributeId="resource:type" Category="resource"
          DataType="string"/>
        </Match>
      </AllOf>
    </AnyOf>
  </Target>
  <Condition>
    <Apply FunctionId="function:integer-at-least-one-member-of">
      <AttributeDesignator AttributeId="resource:patient-id" Category="resource"
      DataType="integer"/>
      <AttributeDesignator AttributeId="subject:patient-ids" Category="subject"
      DataType="integer"/>
    </Apply>
  </Condition>
</Rule>
<Rule Effect="Deny" RuleId="deny-everything-else"/>
</Policy>
```

Fig. 4.13 A simple XACML policy that depicts the permission assignment to appropriate users. Note, the XACML syntax has been simplified to improve readability but this is a fully functional policy document

We also provide a policy-based approach to address the adversary. Let us consider a simple XACML policy that we illustrate in Fig. 4.13. The policy covers access of doctors to the output of heart sensors attached to patients. The policy contains two rules and the first rule that is applicable to the request is evaluated as per the rule combining algorithm attribute. The first rule applies to "get" actions on "heart sensors". This is specified in the target section of the rule. The condition for this rule is that the patient-id attribute associated with the resource (i.e. the name of the patient on which the sensor is attached) must appear in the list of patient-ids associated to the subject (i.e. the list of patients handled by the requesting doctor). If this condition evaluates to false, then the access is denied. The second rule simply returns "deny" if the first rule is not applicable.

If there is no authorization, any doctor can see the sensor-related output of any patient, which in obvious is not an ideal condition. At the very least it jeopardizes patient privacy. To address the authorization issues, XACML policy specifications and their enforcement is necessary. Now we address how our proposed access control architecture deals with the aforementioned adversary scenarios.

- *Scenario 1:* In our proposed access control architecture, the RM verifies the roles of the user based on the attributes recovered from the UAD. If satisfied, the RM consults the appropriate policies with the PMU and then desired capability is issued by the CG. A user cannot claim attributes that have not been registered in the system.
- *Scenario 2:* Recall, that the *Condition Rule* (illustrated in Sect. 3.4.6 of Chap. 3) is written within the conditions specified in the XACML policy. These conditions are checked by the TH upon presenting an access request by the UD. For a capability that can only be used between 9am and 5pm, this will be explicitly mentioned in the condition. The condition will be checked by the TH and access only allowed at the proper times.
- *Scenario 3:* In Sect. 3.6.3 of Chap. 3, we showed that when an access request arrives at the TH, it decrypts the capability using the key shared with the TRR. The TH also obtains the session key (i.e. $K_{UD,TH}$) and decrypts the remainder of the message, and check the authenticity of the UD to whom the key was issued, thus assuring that the correct UD is the one making the request. As the unique UD to whom the capability was issued and the TH are the only entities that know the key $K_{UD,TH}$, only the correct UD would be able to formulate the message.
- *Scenario 4:* In Sect. 3.6.2 of Chap. 3, we discussed that when a capability is generated, the entire information is encrypted under the key the TH shares with the TRR (i.e. K_{TH}). This key is unavailable to the adversary and therefore the capability that is supplied to the TH with the request cannot be modified.
- *Scenario 5:* As illustrated in the above XACML policy specification (cf. Fig. 4.13), if an entity wants to access a resource then the resource must match the identity of a TH listed in the capability. If the entity attempts to use a capability to access a TH not listed in the capability, such access will be denied. Returning to the use case example of Sect. 3.3 of Chap. 3, where an authorized user, e.g. Doctor A, wants to see heart-related medical records of all the patients that are admitted to a particular hospital, not just the patients under his care.

4.6 Summary

This chapter provided the implementation details and performed an evaluation of the achieved results of the access control architecture that is depicted in Chap. 3. We observed that, previous works do not incorporate the advantages of RBAC, ABAC and capabilities to the extent of our proposal. This is especially true when considering capability management and in taking full advantage of the flexibility offered by attributes. Moreover, our proposal extends on previous works that are highly centralized in nature by leveraging the edge intelligence of the IoT through locally evaluating authorization requests by capability evaluation. Importantly, a number of the works cited in Sect. 2.3.3 of Chap. 2, especially those with similar aims to ours, do not provide extensive implementation information.

We evaluated the performance of the proposed architecture using symmetric key cryptography. Our proposed system improves upon the performances (of using asymmetric key cryptography) as it does not depend upon a PKI-based system for authentication. Note that the results in Fig. 4.3 represent a worst-case scenario (i.e. the most time consuming). As discussed in Sect. 3.6 of Chap. 3, if the user already holds a suitable capability then communication with the CMS, and the time represented in Fig. 4.4 is not required. The majority of the time in this case is taken by the decryption operations in the TH, for the capability and access token. Importantly, with a more powerful TH this figure may decrease significantly. Even as it stands, with the TH in our setup implemented on a cheap commodity device, the times are comparable to other proposals as discussed above. Our results also showed that the use of role membership based on attributes is practical and useful while considering the scale of the number of users and the provision of fine-grained access to resources. It can be easily incorporated with symmetric key cryptography with a full-fledged ABAC engine. We argue that a symmetric key based approach is realistic in a context, e.g. a hospital, where pre-registration of the actors is feasible.

We also evaluated the performance of the proposed architecture in the context of asymmetric key cryptography. The performance is comparable with the existing CapBAC architectures e.g. [11, 20]. To the best of our knowledge, those are the two previous works that provide both a conceptual framework and implementation evaluation with results for a CapBAC architecture for the IoT. However, these two proposals do not include the capability generation process, whereas we provide an extensive discussion and implementation for a capability generation process along with its implementation and evaluation. In our case, a capability is generated from XACML policy rules. Note, unlike our approach, proposals [11, 20] used highly optimized ECC and a more powerful microcontroller which contains 128 KB of ROM and 128 KB of RAM. The limiting factor remains signature checking, with the time requirements of the unique features of our system that do not involve signature checking being two orders of magnitude less than state of the art in signature checking employing ECC. While our system still requires such checks, we are not adding significant extra resource requirements in providing flexible and fine-grained access control. the user device and central management system will not be necessary in many cases, as our capabilities can provide access to more than one smart *thing* without loss of security.

In the next chapter (Chap. 5), we plan to investigate the notion of identity and its management for the IoT. This will in turn help to understand the use of attributes for identifying an entity rather than using a predefined concrete identity of that entity, which is immensely important in case of a large-scale and dynamic system like the IoT.

References

1. extensible access control markup language (xacml) version 3.0, http://docs.oasis-open.org/xacml/3.0/xacml-3.0-core-spec-os-en.html (2013) Accessed 02 Nov 2017
2. Mqtt version 3.1.1, https://mqtt.org/. Accessed 02 June 2018
3. J. Dizdarević, F. Carpio, A. Jukan, X. Masip-Bruin, A survey of communication protocols for internet of things and related challenges of fog and cloud computing integration. ACM Comput. Surv. **51**(6), 116:1–116:29 (2019), http://doi.acm.org/10.1145/3292674
4. Z. Shelby, K. Hartke, C. Bormann, The constrained application protocol (coap). Technical Report (2014). http://www.rfc-editor.org/info/rfc7252
5. Mqtt vs. http: which one is the best for iot?, https://medium.com/mqtt-buddy/mqtt-vs-http-which-one-is-the-best-for-iot-c868169b3105. Accessed 21 Feb 2018
6. V. Sarafov, Comparison of iot data protocol overhead. Netw. Arch. Serv. (2018). https://doi.org/10.2313/NET-2018-03-1_02
7. GitHub, "abc.xacml," https://github.com/abc-software/abc.xacml (2016). Accessed 25 May 2017
8. S. Gusmeroli, S. Piccione, D. Rotondi, A capability-based security approach to manage access control in the Internet of Things. Math. Comput. Modell. **58**(5–6), 1189–1205 (2013), http://dx.doi.org/10.1016/j.mcm.2013.02.006
9. nodeMCU, http://www.nodemcu.com/index_en.html. Accessed 20 Apr 2017
10. S. Chandra, S. Paira, S.S. Alam, G. Sanyal, A comparative survey of symmetric and asymmetric key cryptography, in *2014 International Conference on Electronics, Communication and Computational Engineering (ICECCE)* (2014), pp. 83–93
11. J.L. Hernández-Ramos, A.J. Jara, L. Marín, A.F. Skarmeta Gómez, DCapBAC: embedding authorization logic into smart things through ECC optimizations. Int. J. Comput. Math. **93**(2), 345–366 (2016), http://dx.doi.org/10.1080/00207160.2014.915316
12. Cybersecurity pro: Networked medical devices pose huge risks to patient safety, https://www.healthcareitnews.com/news/cybersecurity-pro-networked-medical-devices-pose-huge-risks-patient-safety. Accessed 01 March 2019
13. Sensors facilitate health monitoring, https://www.sensorsmag.com/components/sensors-facilitate-health-monitoring. Accessed 02 March 2019
14. H.L. Tubbs-Cooley, J.P. Cimiotti, J.H. Silber, D.M. Sloane, L.H. Aiken, An observational study of nurse staffing ratios and hospital readmission among children admitted for common conditions, BMJ Qual. Safety **22**(9), 735–742 (2013), https://qualitysafety.bmj.com/content/22/9/735
15. Dhb clinical staffing numbers, https://www.health.govt.nz/our-work/health-workforce/dhb-clinical-staffing-numbers. Accessed 02 March 2019
16. Wireless sensors collect temperature and pressure of bedbound patients, https://www.medicaldevice-network.com/news/wireless-sensors-collect-temperature-pressure-bedbound-patients/. Accessed 02 March 2019
17. K. Chen, S. Zhang, Z. Li, Y. Zhang, Q. Deng, S. Ray, Y. Jin, Internet-of-things security and vulnerabilities: Taxonomy, challenges, and practice. J. Hardware Syst. Sec. **2**(2), 97–110 (2018). https://doi.org/10.1007/s41635-017-0029-7
18. I. Yaqoob, I.A.T. Hashem, A. Ahmed, S.A. Kazmi, C.S. Hong, Internet of things forensics: Recent advances, taxonomy, requirements, and open challenges. Future Gen. Comput. Syst. **92**, 265–275 (2019), http://www.sciencedirect.com/science/article/pii/S0167739X18315644
19. S. Mukherjee, I. Ray, I. Ray, H. Shirazi, T. Ong, M.G. Kahn, Attribute based access control for healthcare resources, in *Proceedings of the 2Nd ACM Workshop on Attribute-Based Access Control*, ser. ABAC '17. New York, NY, USA: ACM (2017), pp. 29–40, http://dx.doi.org/10.1145/3041048.3041055
20. J. Hernandez-Ramos, A. Jara, L. Marın, A. Skarmeta, Distributed capability-based access control for the internet of things. J. Internet Ser. Inf. Sec. **3**(3/4), 1–16 (2013), http://isyou.info/jisis/vol3/no34/jisis-2013-vol3-no34-01.pdf

Chapter 5
Modeling and Management of Identity

Abstract This chapter presents a detailed discussion on modeling and management of identity for the IoT. In a large-scale system like the IoT, it is difficult to predict, in advance, which entities will interact and require access to services and to precisely identify the exact services to which they will seek access. This chapter highlights important questions concerning the nature of identity and identity management for such IoT systems. Further, a formal model of IoT identity covering all its aspects is discussed.

5.1 Introduction

Identity management plays an important role in many application contexts. For instance, e-commerce, e-government, online marketing, just to name a few areas. There are a variety of approaches and their representation for management of these identities and their secure distributions. Identity is often a crucial concern for interoperability (e.g. in surveillance contexts) and privacy (e.g. personal data protection in a healthcare application) [1]. When it comes to the IoT context, identity and its management plays a significant role due to the characteristics of these systems (cf. Chap. 2). Therefore, with the widespread expansion of the IoT, among others, the creation, management and usage of digital identities is one of the prime issues [2]. More and more identities and credentials will be generated, making their management both important and challenging for service providers and users. In the IoT, it is no longer sufficient to simply manage the identity of a person or device that is connected to an application or service. It is more to manage where an entity may possibly be connected to a large number of interconnected heterogeneous *things*.

5.1.1 Problem Description

In Chap. 2, we noted that digital identity is considered as one of the major bases for authenticating and authorizing devices, their management and data flows in an

S. Pal, *Internet of Things and Access Control*, Smart Sensors, Measurement and Instrumentation 37, https://doi.org/10.1007/978-3-030-64998-2_5

online system. Digital identity and identify management have been studied for some time [3]. The majority of previous work, understandably, addresses the issue of digital identity for general computing systems without considering the needs of the IoT. In particular, these identity models typically consider only user identity [4], although some do address service provider identity. In the IoT, identity also needs to capture all the constituent *things*. Identity has been defined and represented in various ways e.g. as globally unique identifiers, a combination of user characteristics [5], a set of attributes of the users [6] or even a set of claims [7], just to name a few approaches. However, when it comes to the IoT, the majority of approaches do not address the issues of identity and its management precisely, keeping in mind the correlation between the dynamic nature and scale of the number of *things*. The issue of unreliable identification of subjects and objects, concerns for security, especially the reliable propagation of sensitive information, emphasises the importance of an appropriate approach to identity (and its management) in the IoT. For instance, Glässer and Vajihollahi [8] discuss identity with different approaches but they do not consider the IoT issues. While Anggorojati et al. [9] take account of IoT issues and their management by incorporating identity-based capabilities and contextual information, they do not examine the issues of IoT identity in a comprehensive manner. We provide a detailed representation of these approaches in Sect. 5.2.

Most studies are concerned with how identities can uniquely identify a particular entity. For example *"a single identity cannot be associated with more than one entity"* [3]. This is true even when identity is based on attributes (e.g. name, age, location, etc.). We argue that such an approach is not sufficiently flexible for a large and highly dynamic system like IoT. When considering issues e.g. policy management and delegation in the IoT, we need to be able to flexibly handle questions of identity. Recall, we already discussed in the previous chapters, that in the IoT, it cannot always be known in advance which users will access which services or devices or which devices will be available at the time when access is requested.

Further, from the above we can observe that there are significant challenges that still need to be addressed for identity management for the IoT. A few frameworks have been proposed (cf. Sect. 5.2.2), and some that have drawn from work in other contexts. However, it has not been demonstrated that they adequately address the particular nature of the IoT, including its scale and heterogeneous context. Moreover, we argue that, identity management for the IoT is not just the use of a unique 'identifier' allocated to each user, device, *thing* or service. It should deal with the issues of attributes and information that can be uniquely used for an access control process that allows a legitimate user to perform an authorized operation. In the IoT, many *things* may conform to the same identity and one identity may refer to a set of *things* performing different operations. Therefore, it is unlikely that representing the identity of each *thing* based on their concrete unique identity will be sufficient. For instance, specifying rules based on the identity of each individual light in a building is not feasible in an IoT context. An alternative could be to represent 'the lights of building E6A'. Any approach should also account for the ability to address collections of entities, and a given collection may be made up of users, services and *things*. This requires dealing with both the scale and dynamic nature of the IoT.

5.1.2 *Contributions*

To address the aforementioned issues of IoT identity and its modeling, in this chapter, we examine the notion of identity from IoT perspective in a systematic and comprehensive manner. In particular, we make the following contributions.

- We study the core concept of digital identify and present a novel idea of IoT identity from the *things* perspectives.
- We propose a formal model of IoT identity and discuss its feasibility in real-world IoT scenarios using practical use case examples.
- We employ attributes for the authorization and authentication of an entity, and it does not depend upon the concrete identities of the entities.

There are a variety of approaches available to identity management [5], including centralized identity management, federated identity management, isolated identity management [10], etc. The choice of such an approach is typically a separate question to the nature of the identities supported. We note that identity and identity management are closely associated, however, the scope of this chapter is limited to identity and does not consider an architecture for identity management. For simplicity, our proposal can easily be adopted to any of the available digital identity management frameworks that we discussed in Sect. 2.4 of Chap. 2.

5.2 Core Concepts

In this section, first we discuss and define identity in detail and then provide a comprehensive analysis of the various representation of identities.

5.2.1 *Identity*

There is no unique definition of identity, it is subjective and depends upon the applied environments. Commonly, the identity of an entity, subject or object, refers to the fact of *who it is*. Usually the identity of an entity is represented as a set of identifiers [5]. The identifiers can be referred to as the characteristic elements of that entity, that are used for identification process. Thus, each identity is reflected by a set of identifiers. Notably, the number of identifiers is greater than the number of identities, which is again larger than the number of entities. Each identity is mapped to one entity, but an entity can have more than one identity mapped to it. Identity can be permanent or temporary depending upon the system's context. The identity has the ability to distinguish between various entities when performing a common action within a specific scope in which they are valid and explicit. To this regard, identity is dynamic and may change upon the context and purpose. Even in a single scope, an entity may have multiple identities, which signifies the fact that the identity is not unique [11].

Identity (and its management) is crucial for the IoT to ensure that only authorized users access the system and the information it contains. In a system where users may have a large number of devices, being able to easily identify them, both uniquely and as groups, is vital in ensuring ease of use, security, privacy and trust. Recall, identity management enables verifying the identity of a legitimate user and the resources that the user is attempting to access [12]. Identity management in the IoT encompasses the creation and usage of digital identities for users, devices, *things* and systems.

Therefore, an identity can be seen as the unique representation of an entity with certain characteristics that can be easily specified using a set of properties. Notably, digital identity differs from the physical identity in terms of its usability and presentation [13]. In a physical identity management process, a person's identity can be their name, age and employee registration number, which uniquely identifies the person within the organization represented by a token e.g. a work identity card. Recall, when we refer to identity for the IoT, we generally refer to the digital identity. A digital identity can be defined as *"a form of identity resulting from the digital codification of identities in a way that is suitable for processing and interpretation by computer systems"* [5].

5.2.2 *Representation of Identity*

Now we provide a comprehensive discussion on different representation of digital identity. We examine the term identity in a broader scope not limited only to the IoT and its associated services and applications. We classify these various characterization based on their representation and properties.

Identity is a Set of 'Partial Identities': Clauß and Köhntopp [14] represent identity as the combination of personal data that are associated to a user (i.e. a human user). The identity of the person is denoted as the set of combinations of a 'partial identity' that are used for certain contexts. Depending upon the situation and the context, a person may be represented by different partial identities. For instance, the partial identity of Alice may vary between her bank and her study-place. In Fig. 5.1, we illustrate the partial identities for Alice in two different contexts. In such, for bank, a partial identity may be composed of name, date of birth and bank account number. Where, for university, the partial identity of Alice may be composed of name, date of birth and her student identity card number. However, even though a user may have a multitude of partial identities, the emphasis here is on linkability and uniqueness, allowing for example, the repeated use of a partial identity to build up a reputation.

Such et al. [15] formalize such partial identities as follows: given a finite set of attributes $A = \{a_1, \dots, a_n\}$ each one with finite domain $V_{a_i} = \{v_1, \dots, v_k\}$, a set of entities E and the entity $e \in E$, a partial identity of the entity e is a vector $I_e = (i_1, \dots, i_n)$, satisfying $i_j \in V_{a_j}$ and $\forall d[d \in E \setminus \{e\} \rightarrow \forall I_d(I_d \neq I_e)]$. Here, V_a is denoted as the value of each attribute a $(a \in A)$ and the set of entity E $(e \in E)$ are

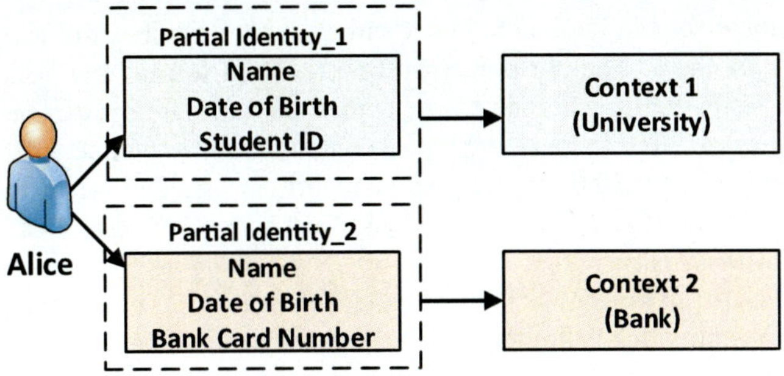

Fig. 5.1 The use of partial identities in different contexts

context dependent. Further, the authors argue that an identity of an entity is composed of many partial identities of the said entity. The formalize identity of the an entity e is: $I_e = \bigcup I_e^j$. Here, identity of entity e, denoted as I_e, is the union of all partial identities (denoted as I_e^j) of e.

Similar to the concept of [15], Ferdous and Poet [16] represent identity as a set of partial identities and they formalize the partial identity and identity as follows: for a domain (d), the partial identity of a user (u), where $u \in U_d$ within d, can be denoted as, $parIdent_d^u = \{(a, v)|a \in A_d, atEntToVal_d(a, u)\ is\ defined\ and\ equals\ v\}$. Where, U_d denotes a set of users, a is a single attribute of u, and v is the value of a. Notation $atEntToVal_d$ is represented as a function of attribute-value pair of an entity in domain d. This function returns the corresponding value of the attribute in domain d and is represented as: $A_d \times U_d \rightarrow AV_d$, where, A_d is a set of attributes. A domain is referred to as an environment in which an entity exists and works. The authors further argue that, if there are n number of attribute-value pairs for a user u in a domain d, then partial identity can also be represented as follows: $parIdent_d^u = \{(a_1, v_1), (a_2, v_2), (a_3, v_3), \ldots, (a_n, v_n)\}$. Finally, the identity of user u is represented as the combination of all partial identities in a domain d, as follows: $Ident_u = \bigcup\{(d, partIdent_d^u)|d \in D\ such\ that\ u \in U_d\}$, where D is a set of domains.

Identity is a Set of Attributes: Some authors directly consider the identity of an entity to be built from attributes without the use of partial identities. For instance, Dabrowski and Pacyna [17] constitute identity as a set of attributes (e.g. name, bank card number, passport number, etc.) belonging to a particular entity that uniquely distinguishes that entity from another. The entity can be a person, a device or a network service that exists in real-life. The authors argue that the identity can be temporary or permanent based on the identity management. Identity management here includes attestation of entity genuineness and trust establishment through the identity authentication. Similar to [17], Gomez-Skarmeta et al. [18] represent identity as a set of attributes that can uniquely identify a person, machine or service that may have many attributes. The authors further propose that each identity is managed by a trusted entity of its corresponding home domain. The home domain is viewed as a domain where each identity is unique.

An attribute set can be denoted as a variable that describes the various characteristics of an entity. The attributes can be presented formally as follows [19]: if $attr_i^e$ represents the i_{th} value of entity e, then the attribute set can be defined as: $ATTR^e = \{attr_1^e, attr_2^e, \ldots, attr_n^e\}$ that contains with n number of attributes of entity e. For instance, the attribute-based identity of student Alice can be represented as follows: $Alice_{student} = \{Student_{id} + Name + Phone\ Number + Date\ of\ Birth + Subject_{code} + Department\}$

Alpár et al. [6] also represent identity as a set of attributes. These attributes form the identity of an entity that is limited within a specific scope. The authors use attribute-based credential technology for a flexible and privacy-preserving authentication of the said entity. The said attributes can be 'identifying' and 'non-identifying', i.e. some attributes hold for a single individual and some attributes hold for many people. For instance, for the former, 'bank account of Alice' identifies the sole holder of the account, where for the former, say attribute 'female' in general does not identify a specific entity [20].

Identity is a Set of Pseudonyms: Bichsel et al. [21] represent identity as a preferred combination of user's credentials and 'pseudonyms' chosen by the user. These combinations are governed by a set of policies that the user must satisfy before accessing a resource. In this case, the pseudonym is the equivalent of the user's public key. The authors posit that, *"unlike public keys of which there is only one for every secret key, however, users can generate an unlimited number of unlinkable pseudonyms for a single secret key. Users can thus be known under different pseudonyms with different verifiers, yet authenticate to all of them using the same secret key"* [21]. However, this still assumes a unique link between pseudonym and user.

As discussed by Hansen et al. in [11], for a user, except for their real-name, anything can be regarded as the pseudonym, even if they belong to hardware or software in the individual's possession. The pseudonym can be represented as an identifier where only the entity that assigned the pseudonym knows the real-life identity behind it. The pseudonym can be self-assigned or defined by a trusted third-party [3]. Importantly, the use of pseudonyms as identifiers helps to protect anonymity in an identity management process.

Identity is a Representation of an Entity: El-Maliki and Seigneur [22] constitute identity as the representation of an entity in a specific context. Chen et al. comprise identity as the representation of an entity within a specific application domain [23]. Similar to [22, 23], Jorns et al. [24] also construct identity as the representation of an entity in a specific context or dedicated application domains.

L'Amrani et al. [25] represent identity as an entity that is combined with context and characteristics. The entity can be a person, organization or resource. The context is the environment for which the identity is defined, and if the context changes the associated identity can change. The characteristics can be defined implicitly (e.g. manually) or inherits naturally (e.g. fingerprints, voice, biometrics, etc.).

Identity is an Individual Characteristic: Jøsang et al. [5] discuss identity of a person consisting of the individual characteristics by which that person is recognized. For instance, Alice in an organization can be identified as the unique combination of name, address, nationality and passport number. This further argues the definition of [6] that represents identity as the set of user's attributes. Angin et al. [26] also constructs identity as a set of unique characteristics of an entity that is used to identify the entity with identifier.

Identity is a Set of Claims: Cameron [7] represents identity as a set of 'claims' made by one digital subject about itself or another digital subject. The claims can be any identifier or a set of identifiers. For instance, for a student, claims can be the student number provided by the university or a set of identifies e.g. name, address, date of birth and nationality. This further argues the use of attributes for the identity as in [17]. Where these claims can be denoted as unique (or a set of) user's attributes for a specific purpose, in this case to identify a student within the university.

Formally a claim can be seen as a statement about a user, similar to an attribute assertion in Security Assertion Markup Language (SAML) 2.0, expressed and signed by the identity provider. However, for obtaining such claims, the users need to be authenticated themselves to the identity provider [27].

Identity is a Statement: Fongen [28] refers to identity as a 'statement', similar to a public key certificate in the sense that it attests a binding between a public key and the identity information of the 'owner' of the private key. The identity statement is issued by an identity provider similar to a Certificate Authority (CA) like service. Formally, the identity statement of principal x signed by the identity provider of a Communities of Interests (COI) a can be denoted as $(Id_x)_a$ and is represented as follows:

$(Id_x)_a = Name_x + PublicKey_x + Attributes_x + Timestamp + Serialnumber + Signature_a,$ *where,* $Attributes_x$ is the set of name-value pairs and $Signature_a$ is the signature of the identity provider of COI a. The COI represents the group of similar interests within a domain. For each COI there is one unique identity provider. Here, the domain is a unit of trust, authority and administration with some specific properties [29].

Identity is a Set of Credentials: Cao and Yang [30] consider identity as the representation, proofs and credentials of user entity which should allow the user the desired services and applications. This is bounded by a specific context. The credential could be the digital certificates (e.g. X.509), One Time Password (OTP) or Personal Identification Number (PIN).

Identity is an Instrument: Torres et al. [31] represent identity as an instrument that can be used for providing information to a system about itself. It is associated with an entity or generally formed by a unique identifier. The authors argue that the identity is a virtual concept and does not exist in real-life.

A number of the authors mentioned above highlight the importance of supporting privacy in the context of a shift to a digital society. To this end it is worth considering approaches to identity management that focus on privacy. For example, Camenisch et al. propose a privacy-aware identity management system 'PRIME' (PRivacy and Identity Management in Europe) which focuses on giving users control of their personal data [11]. The system employs cryptographic mechanisms to provide secure and anonymous communication and to support various forms of identity including pseudonyms (i.e. conceal a user's or system's identity). The primary goal of 'PRIME' is the maintaining of the users privacy and accountability in an electronic transaction. The system architecture includes users, service providers and certification authorities.

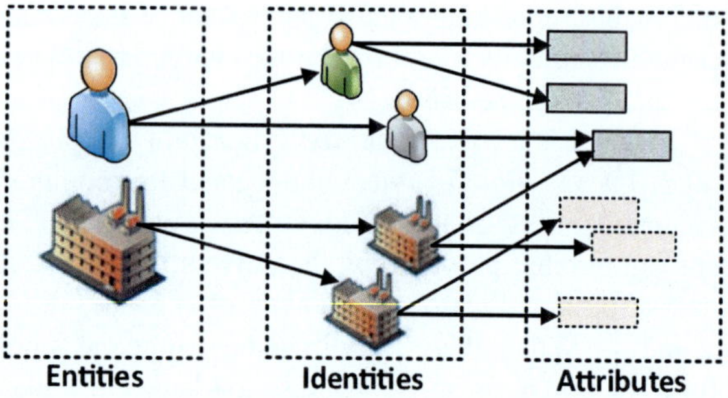

Fig. 5.2 Standard relationship between entities, identities and attributes

Users are the entities who are requesting service from the service providers. Service providers are the computing systems that provide services to the users by means of transactions. Finally, a certification authority is a special type of service provider which can issues certificates that are signed digitally by them. While promising, its applicability to the IoT is uncertain as it relies on a centralized database that holds certificates and uncertified data of entities, as well as policies. Even if this proves unsuitable for the IoT, the emphasis on privacy is notable.

In summary, what these approaches typically have in common is an emphasis on being able to link identities (even partial entities and pseudonyms) uniquely to a particular entity, as can be seen in Fig. 5.2 above, and also Fig. 1 from [14]. This is not sufficient for a general IoT context. Even grouping users together (for example in groups and roles) still requires unique identification of those users. It is difficult to predict, in a general IoT situation exactly which users may interact with which devices. Therefore unique identity is not the ideal basis for identities from a policy management point of view. Similarly, identifying the targets (e.g. connected LED lighting systems) of policy by (for example) unique device identity can result in inflexible and overly-detailed policy expressions. Many of the existing solutions rely on attributes, or related concepts, but still insist upon uniquely identifying entities.

We build upon and extend the previous approaches and their use of attributes, where users are identified by sets of attributes (e.g. name, age, location, etc.). Significantly, in ABAC, access control permissions are assigned based on the policies that are governed by the attributes. These attributes can be seen as the properties that describe specific features of users, resources, contexts and conditions. The attributes of the user and those of the resource together determine the set of operations (based on the policies) that can be performed in a specific context. However, in Sect. 3.2.2.1 of Chap. 3, we discussed that in ABAC, while multiple users may be referred to by an attribute set in a policy (i.e. using a set of rules), typically there is no concept of such a set constituting an identity or being able to be referred to as a distinct grouping (so, for example, being able to be used in multiple policy expressions), nor is the idea extended to the targets of the policy.

5.3 An Approach to IoT Identity

In this section, we present our approach to IoT identity. First, we discuss the notion of IoT identity. Second, we provide the requirements and properties that can be considered when modeling an IoT identity. Third, we illustrate our approach for IoT identity based on the *thing's* perspective. Finally, we give a formal specification of the model.

5.3.1 IoT Identity

Now we examine identity for the IoT from the lessons learned. Some contributions have discussed IoT identity and its management. For instance, Chen et al. [23] present an identity management framework for IoT. In this, identity is represented as a set of general information of an entity e.g. identifier, access information, etc. Thus, in this case, identity is predefined and uniquely assigned to each entity. Fongen [32] presents a view of identity management and integrity protection in the IoT. In this, the concept of identity is denoted by the unique device identity. Sarma and Girão [33] present a discussion on identity in the IoT with 'identinet' and 'digital shadow'. The identinet represents each end-point (e.g. user, device etc.) with a unique identity. Digital shadow represents the information that an entity uses. Once again, this approach leads identity to a device level. Further, despite the flexibility promised by these concepts, issues of scalability, interoperability and identity management in their implementation in a heterogeneous IoT system are not addressed.

Similar to [33], Thuan et al. [34] present an identity management system for the IoT that is user centric and allows device authentication and authorization based on unique user identity. Trnka and Cerny [35] and Santos et al. [36] discuss identity management for a collection of devices in an IoT system. Mahalle et al. identify a number of challenges for identify management in the IoT [37]. These include the preservation of privacy of identity and the need for single sign-on (SSO) facilities to handle the scale of interaction in the IoT. While they outline a framework for identity management in the IoT, it is only briefly described. Once again, in these approaches, identity is represented as the unique user or device identity.

A major challenge in the IoT is device identification i.e. what the device is, and authentication i.e. is the device the one it claims to be. Cryptographic protocols can be a solution, however, the resource-constrained nature and scale of IoT systems makes it infeasible to apply them directly to the IoT. To address this issue, recent proposals e.g. [38, 39] argue that the device's behavioural fingerprinting can be used to solve both of these problems effectively. Device behavioural fingerprints are built up by observation of the device, e.g. via its network traffic and effectively consist of dynamic attributes of the device. Machine learning techniques are applied to determine whether the device is behaving within acceptable norms. The major aim of behavioural fingerprinting is to detect any malicious activity by periodically observing and validating against an ideal behavioural profile of the device.

Besides device fingerprinting, the majority of the above contributions rely on unique identities, typically based on unique device characteristics. Device fingerprinting has not, so far, been used for wider identity purposes, but does depend on more general attributes.

5.3.2 Requirements and Considerations

We notice that, most of the aforementioned proposals (cf. Sect. 5.2.2) for IoT identity build upon a concrete identity-definition (i.e. knowing the subjects and objects in advance), which is unsuitable for a large and highly dynamic system like IoT. We argue that the IoT identity should support the characteristics of an IoT system to provide a flexible and fine-grained identity modeling [40, 41], for instance, identity construction must consider the characteristics of an IoT system (cf. Sect. 2.1.2 of Chap. 2). For a quick recap, we highlight them as follows:

- The scale of the system, where potentially billions of *things* are interconnected with one another.
- The highly dynamic nature of the system, where interactions between the *things* may be flitting and happen only once.
- The heterogeneity of devices, systems and applications. This may contain various operating platforms, network technologies and communication protocols.
- The resource-constrained nature of IoT devices, that are limited to processing speed, memory capability or even battery power.
- The high number of management domains where integration of services and application can have a crucial issue, just to name a few characteristics.

However, the design choice depends fully upon the specific environment, system requirement and designer's preferences. Given the dynamic base of such IoT systems, we observe that identity management should react to the change in entity attributes over time and context. We propose the following properties which can be considered when modeling an IoT identity.

- An identity consists of a set of attributes, and their values, the members of the set varying in the different contexts of an IoT system (e.g. different service provision or different layers of the architecture).
- An identity can refer to a collection of *things* that function in a specific context (e.g. location awareness).
- An identity should have a purpose e.g. to provide a set of services (e.g. to supply data from physical objects).
- An identity should be treated uniformly across heterogeneous platforms (e.g. identity of an entity can easily be noted across the domains).

- The values of attributes in an identity can be dynamic and these can change with the membership of identity collection (e.g. attribute-based identity credentials).

This results in a situation where entities may share identities on a dynamic basis. This remains under the control of the system authority that creates the identity, as they define the attributes and values that constitute the identity. As the interactions with the IoT may be spontaneous, short-lived and, in quantity, large-scale, it is easier to manage them on the basis of attribute sets (i.e. partial identities) which define the intended participants (e.g. users, devices, etc.) in the interactions, rather than the concrete identities. Towards this, in the next section, we discuss our approach for IoT identity.

5.3.3 Our Approach: Things-Centric Identity

We argue that the scope and nature of an IoT system mean that insisting on definitive, unique, identity in every case is overly restrictive. While in some circumstances such unique identification will be required, in other cases less defined identities will suffice for the needs of application functionality and policy specification. Unique identities will not always be known in advance (for example, which customers may enter a shop or purchase a movie ticket), however partial identities (e.g. age) may be able to be defined in advance. We further contend that such partial identities are best represented as sets of attributes. This is in accord with the majority of previous work on identity, as summarized in Sect. 5.2.2, however we differ from previous work in moving away from an insistence on unique identification of users as the core purpose of identity.

Identity functions within a context, e.g. an application domain. Identity itself consists of the total sum of the attributes of the entity within that domain and can be viewed as the set of partial identities of the entity (or *thing*) within that context. The *thing* can, for example, be a device, an application, a human user or even an organization. Partial identities consist of one or more attributes. In other words, an attribute (or a set of attributes) identifies and distinguishes a *thing* within a specific context from other *things* to a greater or lesser extent. This can be seen as *things*-centric identity for the IoT. Note that identity then exists within a certain context and *things* will likely have different identities (and partial identities) within different contexts. This is a deliberately functional approach to identity, identity for us only matters when it serves to identify a *thing* within a particular context. A *thing*'s identity consists of all the attributes it possesses which are applicable within a particular context. A partial identity is then a subset of those attributes and again partial identities only have meaning within the particular context(s) in which those attributes are recognized. Just as *things* may have more than one partial identity within a context, more than one *thing* may share the same partial identity (cf. Fig. 5.3). For example, multiple people may share the partial identity (age = '47', credit_rating = 'good'). We recognize that the chosen attribute set of a *thing* may not be sufficient to uniquely

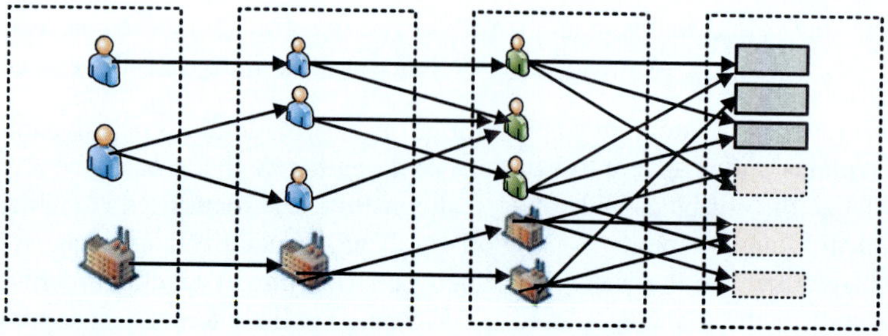

Fig. 5.3 The proposed vision of IoT identity and its relations with entities, partial identities and attributes

identify it within a certain context, however not all use-cases require unique identity (cf. Sect. 5.4).

5.3.4 *Use of Attributes*

In Chap. 3, we discussed that the *things* in an IoT system can be represented by a set of attributes rather than depending upon their concrete identity (for example, all the lights in a dwelling rather than the individual low-level addresses of the devices controlling such lights). We also argued that the use of attributes can reduce the overhead on the system by avoiding the need to store and specify policies based on the identity of each entity. In other words, the authentication of an entity should not be dependent upon the concrete identity of the entity. As noted above, our design uses attributes. Attributes are a well-known mechanism for use with security and access control. Attributes describe an aspect of an entity or context, e.g. age, job title, location or function.

In our proposal, attributes are used for the authorization and authentication of an entity rather than depending upon their concrete identities. In essence, the attributes can be used to group *things* together. For instance, an attribute assigned to 'all lights in my house'. When policies are expressed, they can refer to the attributes of the entities, rather than their individual identities. This is, in particular, useful when composing policies for all members of the group. With such an approach, users can easily express policies which, for example, give all guests access to all lights in their house, rather than having to laboriously specify individual policies for each potential guest and light. Users of the system do not have to remember who all their 'guests' are (or, more precisely, who all the other users are to which they have allocated the attribute 'guest') but can easily write policies specifying the rights provided to 'guests'. This approach allows a more flexible view of identity, reflecting the dynamic nature of the IoT. By conceptualizing identity based around attributes, the linking of entities to identities can dynamically change as their associated attribute values change.

The use of attributes thus provides a powerful method of specifying access policies in a flexible and fine-grained way that is particularly useful in an IoT system. In the previous chapters (Chaps. 3 and 4), we show how using attributes, we can flexibly define role-membership, which in turn specifies the available permissions. This reduces the number of policies that must be created by allowing a single attribute expression to provide access to multiple resources. We argue that the same attributes may have different values in different contexts. For instance, let us consider the attribute 'qualified' which may exists in two contexts e.g. 'a taxi driver' and 'a plane pilot'. In other words, in this case, the attribute 'qualified' may have the value 'true' in the context of a taxi driver but 'not true' in case of a plane pilot.

5.3.5 A Formal Specification

Now we illustrate a formal model of our proposal. In specifying a formal model of IoT identity we follow the approaches discussed in [15, 16, 27]. The model has the following components: E, Cnt, A, AV, PId and Id (sets of entity, sets of context, sets of attribute, sets of attribute values, sets of partial identity and sets of identity respectively). Where, $e \in E$, $cnt \in Cnt$ and $a \in A_{cnt}$. We can formally represent an IoT identity as following tuple: $IoT\_Identiy\_Model = \{E, Cnt, A, AV, PId, Id\}$, where:

Definition 5.1 (*Entity*) The set of entities is represented by E. It is a *thing*, which could be, for example, a user, device, services and applications. An entity may be either an individual or an organization that has a distinctive existence in a physical or logical sense. Note, E is the set of entities. It can be denoted as follows:

$$E = \{e_1, e_2, \ldots, e_{n-1}, e_n\}$$

For instance, consider, E denotes the set of all entities. The sets of users are defined by U, sets of organizations are defined by O and sets of devices are given by D, then the union of all sets constitute the entire set of entities. This can be seen as:

$$E = U \cup O \cup D, \text{ where,}$$

$$U \cap (O \cup D) = \emptyset$$
$$O \cap D = \emptyset$$

Definition 5.2 (*Context*) The set of contexts is represented by Cnt. It can be defined as a specific application domain where an entity exists and operates. In a context an entity is represented and identified by its attributes. This identification may or may not be unique. Context can be of different types e.g. personal context, place context, time context, etc. A context set is composed of a set of context elements, as follows:

$$Cnt = \{C_1, C_2, C_3, \ldots, C_{n-1}, C_n\}, \ (n \in N)$$

Each context element can be composed of various attributes and the attribute values. This can be seen as: $Context\ Element : \{attribute\_name, attribute\_value\}$, for example, $\{organization, coordinate\}$ represents a location context. The context information plays an important role for making an access decision, as it can hold the identity of an entity.

Definition 5.3 (*Attribute*) The set of attributes is represented by A. An attribute is a variable associated with an entity within a given context. It is a distinct, measurable property of the entity in a specific context and associated with a value that is used to identify (not necessarily uniquely) the entity within the context. All entities must have at least one attribute that can be understood within a certain context for that entity to have an identity within that context. Note, in this chapter, when we refer to attributes we are referring to the attributes of *things*. An attribute set can be denoted as follows:

$$A = \{a_1, a_2, \ldots, a_{n-1}, a_n\}$$

Note that the set of attributes in different contexts may overlap. For instance, for two different contexts cnt_1 and cnt_2 (where, $cnt_1, cnt_2 \in Cnt$), the following does not automatically hold: $A_{cnt_1} \cap A_{cnt_2} = \emptyset$.

Definition 5.4 (*Attribute Value*) An attribute's value can be generated by the entity or can be provided by a third party, and associated with a data type that defines the values an attribute can take. How this is done is out of the scope of this chapter. Note, AV is the set of attribute values. It can be denoted as follows:

$$AV = \{av_1, av_2, \ldots, av_{n-1}, av_n\}$$

Note that the values of the attributes are dependent upon the context. In such, $AV_{cnt} \subseteq AV$ is the attribute values in context cnt $(cnt \in Cnt)$.

Definition 5.5 (*Partial Identity*) It is represented by PId. A partial identity of an entity in a given context is a subset of the attributes and their values that are associated with that entity in that context. A partial identity may correspond to a number of entities. That is a given partial identity may not uniquely identify an entity. This allows multiple entities to be addressed by the use of a single partial identity. In other words, a partial identity may not be sufficient for all application functions with a context. For example, given a context where all entities have the attributes 'age' and 'credit_rating' then a partial identity within this context consisting only of 'age' may not be sufficient for all functions within the context. If the attributes that make up a partial identity are all valid in more than one context then that partial identity is valid in all such contexts. It can be denoted as follows: For an entity e ($e \in E$), in a given context cnt ($cnt \in Cnt$), the partial identity of e can be denoted as follows:

$$PId^{e}_{cnt} = \{(a, e)|a \in A, (av, e)|e \in E\}$$

Definition 5.6 (*Identity*) It is represented by Id. The identity of an entity within a given context can be represented as the union of all its partial identities within that context. An identity must uniquely identify the entity within the context. For an entity e ($e \in E$), in a given context cnt, the identity of e can be denoted as follows:

$$Id^{e}_{cnt} = \cup\{PId^{e}_{cnt}|e \in E, cnt \in Cnt\}$$

Partial identities can be used to identify groups of related entities, for example by referring to them by such a partial identity in the specification of policies. We explicitly allow for a partial identity of two or more separate entities in one context to be the same. While the 'linkability problem' [15] may occur for partial identities, it does not apply to identity under our model. Our intention is to create an identity model that serves application and policy specification needs and these do not always require unique identity specification. We present the following relationship mappings:

- $AE : A \times E$, a many to many attributes to entity assignment relation. This mapping specifies all the attributes that each entity possesses.
- $ACnt : A \times Cnt$, a many to many attributes to context assignment relation. This mapping specifies the corresponding attributes that an entity may possess within that context. An entity must possess at least one of these attributes to have an identity within that context.
- $APId : A \times PId$, a many to many attribute to partial identity assignment relation. This mapping specifies the corresponding attributes that constitute the partial identity of an entity in a given context.
- $AId : A \times Id$, a many to many attributes to identity assignment relation. This mapping specifies the corresponding attributes that constitute the identity of an entity in a given context.
- $PIdId : PId \times Id$, a many to many partial identities to identity assignment relation. This mapping specifies the corresponding partial identities that constitute the identity of an entity in a given context.
- $EId : E \times Id$, a many to many entity to identity assignment relation. This mapping specifies that an entity should hold at least one identity in a given context.
- $(AAV)_{cnt} : (A \times AV)_{cnt}$, a many to many attributes to attribute-value pairs assignment relation within a particular context. This mapping specifies that attributes to a specific context are matched to a specific value.

5.4 Use-Case Examples

In this section, we enhance the use-case example of an IoT-enabled smart health care system that we discussed in Sect. 3.3 of Chap. 3. Our intention is to outline some specific instances to demonstrate the use of our proposed model of identity for large-scale IoT systems. To demonstrate this, we choose three distinct examples that combine end-device, user and, user and end-device point of view. Our choice of these three cases indicatively covers the users, devices and their associated applications and services present in an IoT context.

Scenario 1—End Device's Partial Identity: Suppose, Alice is a doctor treating patient Bob. Bob may have many IoT-enabled sensors attached providing real-time information about Bob's condition to Alice and other staff (e.g. blood pressure, heart rate, body temperature and blood glucose level). As Bob's condition evolves further sensors may need to be activated. Rather than Alice requiring separate credentials for each sensor, the credential that allows Alice access to Bob's sensors may simply specify access to sensors with the attribute '*Attached_to_Bob*' or similar. Note, this can be achieved in our design by using a capability template and *Parameterization Rule* that creates a capability allowing access to sensors attached to Bob defined by the appropriate attributes (cf. Chap. 3). Then, as new sensors are given that attribute and brought into service Alice has immediate access. This will also apply if one sensor breaks and needs to be replaced. This avoids the need to deal with unique sensor identity. In this case, we address the identity of the sensors that are attached to Bob. It represents the sensors' partial identity by which they can be referred to. However, in this case, Alice can be uniquely identified by a full identity.

Scenario 2—User's Partial Identity: Refer back to the example discussed in Sect. 3.4.6 of Chap. 3, there are applications in the IoT where service providers wish to make services available to groups of users without individually, and uniquely, identifying each user. For example, a shop may make specials offers and discounts to particular customer segments, identifying them by ranges of such attributes as age, address and interests. Sets of these attributes would form partial identities without uniquely identifying the users, e.g. age, suburb in which they live, previous purchases in shop, etc. As commercial entities usually wish to attract new customers and not limit their customer pool, unique identification does not serve their purpose in this case. In such cases, the targets of the policy could be definitively identified. For instance, the '*Role – Membership*' based on age and suburb of a resident can be seen as a partial identity according to the partial identity definition expressed using the attribute definition discussed above.

Scenario 3—User's and End Device's Partial Identities: Now consider the following use case, where many doctors may have access to the sensors of a patient. It is inefficient to individually refer to each of the doctors, especially as the doctors treating the patient may change over time. Instead, we can identify the doctors treating the patient by giving all of them an attribute '*Doctor_of_Bob*'. In this case, a partial identity is formed from the attribute '*Doctor_of_Bob*' and possibly other attributes.

These partial identities do not need to uniquely identify a single doctor. This allows many doctors to access the multiple sensors attached to the patient.

Similarly, suppose, in a building the administration wished to give certain staff members (e.g. accounting staff) the ability to control all the lights on the third floor. This could be achieved by giving each of lights attributes that identified them as lights and as being placed on that floor. They may also have further attributes, e.g. the '*type_of_light*' and the '*room*' in which they are located. However, credentials given to the users for the purpose outlined only need to specify that the policy targets have the partial identity of being lights on the third floor. In this case, many lights share the same partial identity, simplifying credential management. The partial identity of the security staff members includes the attribute 'security-staff'.

5.5 Summary

This chapter examined various available approaches and outlined the foundations for building a formal model of IoT identity. Due to the characteristics of such systems, it is difficult to predict, in advance, which entities will interact and require access to services from which entity. This introduces an important question on how to precisely identify the exact services to which they will seek access. An identity can be regarded as a representation of an entity that is used for a specific application context to uniquely define an element (e.g. user, device, etc.). Identity can be temporary or permanent. The characteristics of the element can be called an 'identifier' that helps in the 'identification process'. Identity management can involve authentication, authorization, access control and policy enforcement both physically or digitally.

Several proposals have specified identity based on 'a set of partial identities', 'claims' or even the 'process of representing an entity'. It is important to note that the identity of an entity in one community can be different to their identity in another community. For example, an organization holds certain individual characteristics (e.g. name, address, location of their office, post holds, etc.) by which a person can be recognized uniquely. However, the same person may have a different identity for their online banking account. In other words, the identity of an entity (the person in this case) is valid under the scope of some specific context. Furthermore, identities are dynamic in nature. For example, the age of a person changes and perhaps their role in an organization changes over time. Therefore, we argue that the identity management should react to the change in entity attributes over time.

We have investigated various representation of digital identities and its suitability when addressing identity in an IoT context. This lead us to build an identity model for the IoT. Our proposed model of IoT identity is based on *attributes*, rather than depending upon a concrete identity of an entity. For example, users may wish their friends or family to be able to access certain devices or their output. By allocating the attribute 'friend' to a group of users, policies can be formulated based on that attribute. The system can record who has been allocated that attribute allowing the user to easily comprehend policies and avoiding the need to ensure that every friend is included in policy specification. This ease of use of policies will assist in giving them both actual control and a sense of control. Redundancy will also be catered for, as redundant devices can be referred to by a common set of attributes.

In the next chapter (Chap. 6), we plan to enhance the notion of identity that we discussed in this chapter to a practical testbed to investigate delegation issues (i.e. transferring of access rights from one entity to another) in an IoT context. We further intend to show the use of attributes (to composite an identity) and its flexible and ease use to an access right delegation.

References

1. B. Chao, L. Yang, A survey of identity management technology, in *2010 IEEE International Conference on Information Theory and Information Security* (2010), pp. 287–293
2. A. Al-Gburi, A. Al-Hasnawi, L. Lilien, Differentiating security from privacy in internet of things: a survey of selected threats and controls, in *Computer and Network Security Essentials*, ed. by K. Daimi (Springer International Publishing, Berlin, 2018), pp. 153–172. http://dx.doi.org/10.1007/978-3-319-58424-9_9
3. A. Jøsang, S. Pope, User centric identity management (2005), https://link.springer.com/content/pdf/10.1007%2F978-3-642-23300-5_1.pdf
4. L.J. Camp, Digital identity. IEEE Technol. Soc. Mag. **23**(3), 34–41 (2004). http://dx.doi.org/10.1109/mtas.2004.1337889
5. A. Jøsang, J. Fabre, B. Hay, J. Dalziel, S. Pope, Trust requirements in identity management, in *Proceedings of the 2005 Australasian Workshop on Grid Computing and e-Research - Volume 44*, ser. ACSW Frontiers '05 (Australian Computer Society, Inc., Darlinghurst, Australia, 2005), pp. 99–108, http://portal.acm.org/citation.cfm?id=1082305
6. G. Alpár, L. Batina, L. Batten, V. Moonsamy, A. Krasnova, A. Guellier, I. Natgunanathan, New directions in IoT privacy using attribute-based authentication, in *Proceedings of the ACM International Conference on Computing Frontiers*, ser. CF '16 (ACM, New York, NY, USA, 2016), pp. 461–466. http://dx.doi.org/10.1145/2903150.2911710
7. K. Cameron, The laws of identity. Technical Report (2005), https://msdn.microsoft.com/en-us/library/ms996456.aspx
8. U. Glässer, M. Vajihollahi, Identity management architecture, in *Security Informatics*, ed. by C.C. Yang, M. Chau, J.-H. Wang, H. Chen. Annals of Information Systems, vol. 9 (Springer US, 2010), pp. 97–116. http://dx.doi.org/10.1007/978-1-4419-1325-8_6
9. B. Anggorojati, P.N. Mahalle, N.R. Prasad, R. Prasad, Capability-based access control delegation model on the federated IoT network, in *2012 15th International Symposium on Wireless Personal Multimedia Communications (WPMC)* (IEEE, 2012), pp. 604–608, http://ieeexplore.ieee.org/xpls/abs_all.jsp?arnumber=6398784
10. T. El Maliki, J. Seigneur, A survey of user-centric identity management technologies, in *The International Conference on Emerging Security Information, Systems, and Technologies (SECUREWARE 2007)* (2007), pp. 12–17
11. J. Camenisch, A. Shelat, D. Sommer, S.F. Hübner, M. Hansen, H. Krasemann, G. Lacoste, R. Leenes, J. Tseng, Privacy and identity management for everyone, in *Proceedings of the 2005 Workshop on Digital Identity Management*, ser. DIM '05 (ACM, New York, NY, USA, 2005), pp. 20–27. http://dx.doi.org/10.1145/1102486.1102491
12. A. Ouaddah, H. Mousannif, A. Abou Elkalam, A. Ait Ouahman, Access control in the internet of things: big challenges and new opportunities. Comput. Netw. **112**, 237–262 (2017). http://dx.doi.org/10.1016/j.comnet.2016.11.007
13. M. Hansen, A. Schwartz, A. Cooper, Privacy and identity management. IEEE Secur. & Privacy Mag. **6**(2), 38–45 (2008). http://dx.doi.org/10.1109/msp.2008.41

14. S. Clauß, M. Köhntopp, Identity management and its support of multilateral security. Comput. Netw. **37**(2), 205–219 (2001). http://dx.doi.org/10.1016/s1389-1286(01)00217-1
15. J.M. Such, A. Espinosa, A. Garcia-Fornes, V. Botti, Partial identities as a foundation for trust and reputation. Eng. Appl. Artif. Intell. **24**(7), 1128–1136 (2011). http://dx.doi.org/10.1016/j.engappai.2011.06.008
16. M.S. Ferdous, R. Poet, Formalising identity management protocols, in *2016 14th Annual Conference on Privacy, Security and Trust (PST)* (IEEE, 2016), pp. 137–146. http://dx.doi.org/10.1109/pst.2016.7906948
17. M. Dabrowski, P. Pacyna, Generic and complete three-level identity management model, in *2008 Second International Conference on Emerging Security Information, Systems and Technologies* (IEEE, 2008), pp. 232–237. http://dx.doi.org/10.1109/securware.2008.18
18. A.F. Gomez Skarmeta, P.M. Julia, J. Girao, A. Sarma, Identity based architecture for secure communication in future internet, in *Proceedings of the 6th ACM Workshop on Digital Identity Management*, ser. DIM '10 (ACM, New York, NY, USA, 2010), pp. 45–48. http://dx.doi.org/10.1145/1866855.1866866
19. M. Ed-Daibouni, A. Lebbat, S. Tallal, H. Medromi, A formal specification approach of privacy-aware attribute based access control (Pa-ABAC) model for cloud computing, in *2016 Third International Conference on Systems of Collaboration (SysCo)* (IEEE, 2016), pp. 1–5. http://dx.doi.org/10.1109/sysco.2016.7831324
20. G. Alpár, B. Jacobs, Credential design in attribute-based identity management (2013), http://citeseerx.ist.psu.edu/viewdoc/summary?doi=10.1.1.413.2571
21. P. Bichsel, J. Camenisch, M. Dubovitskaya, R. Enderlein, S. Krenn, I. Krontiris, A. Lehmann, G. Neven, C. Paquin, F.-S. Preiss, K. Rannenberg, A. Sabouri, An architecture for privacy-ABCs, in *Attribute-based Credentials for Trust*, ed. by K. Rannenberg, J. Camenisch, A. Sabouri (Springer International Publishing, Berlin, 2015), pp. 11–78. http://dx.doi.org/10.1007/978-3-319-14439-9_2
22. T. El Maliki, J.-M. Seigneur, A survey of user-centric identity management technologies, in *The International Conference on Emerging Security Information, Systems, and Technologies (SECUREWARE 2007)* (IEEE, 2007), pp. 12–17. http://dx.doi.org/10.1109/secureware.2007.4385303
23. J. Chen, Y. Liu, Y. Chai, An identity management framework for internet of things, in *2015 IEEE 12th International Conference on e-Business Engineering* (IEEE, 2015), pp. 360–364. http://dx.doi.org/10.1109/icebe.2015.67
24. O. Jorns, G. Quirchmayr, O. Jung, A privacy enhancing mechanism based on pseudonyms for identity protection in location-based services, https://dl.acm.org/citation.cfm?id=1274547
25. H. L'Amrani, B.E. Berroukech, Y. El Bouzekri El Idrissi, R. Ajhoun, Identity management systems: laws of identity for models 7 evaluation, in *2016 4th IEEE International Colloquium on Information Science and Technology (CiSt)* (IEEE, 2016), pp. 736–740. http://dx.doi.org/10.1109/cist.2016.7804984
26. P. Angin, B. Bhargava, R. Ranchal, N. Singh, M. Linderman, L. B. Othmane, L. Lilien, An entity-centric approach for privacy and identity management in cloud computing, in *2010 29th IEEE Symposium on Reliable Distributed Systems* (IEEE, 2010), pp. 177–183. http://dx.doi.org/10.1109/srds.2010.28
27. G. Alpár, J.-H. Hoepman, J. Siljee, The identity crisis. Security, Privacy and Usability Issues in Identity Management (2011). arXiv:1101.0427
28. A. Fongen, Architecture patterns for a ubiquitous identity management system. ICONS **2011**, 66–71 (2011)
29. A. Fongen, Identity management without revocation, in *2010 Fourth International Conference on Emerging Security Information, Systems and Technologies* (IEEE, 2010), pp. 75–81. http://dx.doi.org/10.1109/securware.2010.20
30. Y. Cao, L. Yang, A survey of identity management technology, in *2010 IEEE International Conference on Information Theory and Information Security* (IEEE, 2010), pp. 287–293. http://dx.doi.org/10.1109/icitis.2010.5689468

31. J. Torres, M. Nogueira, G. Pujolle, A survey on identity management for the future network. IEEE Commun. Surv. & Tutor. **15**(2), 787–802 (2013). http://dx.doi.org/10.1109/surv.2012.072412.00129
32. A. Fongen, Identity management and integrity protection in the internet of things, in *2012 Third International Conference on Emerging Security Technologies* (IEEE, 2012), pp. 111–114. http://dx.doi.org/10.1109/est.2012.15
33. A. Sarma, J.a. Girão, Identities in the future internet of things. Wireless Personal Commun. **49**(3), 353–363 (2009). http://dx.doi.org/10.1007/s11277-009-9697-0
34. D. van Thuan, P. Butkus, D. van Thanh, A user centric identity management for internet of things, in *2014 International Conference on IT Convergence and Security (ICITCS)* (IEEE, 2014), pp. 1–4. http://dx.doi.org/10.1109/icitcs.2014.7021724
35. M. Trnka, T. Cerny, Identity management of devices in internet of things environment, in *2016 6th International Conference on IT Convergence and Security (ICITCS)* (IEEE, 2016), pp. 1–4. http://dx.doi.org/10.1109/icitcs.2016.7740343
36. B. Santos, V.T. Do, B. Feng, T.v. Do, Identity federation for cellular internet of things, in *Proceedings of the 2018 7th International Conference on Software and Computer Applications*, ser. ICSCA 2018 (ACM, New York, NY, USA, 2018), pp. 223–228. http://doi.acm.org/10.1145/3185089.3185132
37. P. Mahalle, S. Babar, N. Prasad, R. Prasad, Identity management framework towards internet of things (IoT): roadmap and key challenges, in *Recent Trends in Network Security and Applications*, ed. by N. Meghanathan, S. Boumerdassi, N. Chaki, D. Nagamalai. Communications in Computer and Information Science (Chap. 43), vol. 89 (Springer, Berlin, 2010), pp. 430–439. http://dx.doi.org/10.1007/978-3-642-14478-3_43
38. IoTSense: Behavioral Fingerprinting of IoT Devices (2018). arXiv:1804.03852.pdf
39. M. Miettinen, S. Marchal, I. Hafeez, N. Asokan, A.-R. Sadeghi, S. Tarkoma, IoT sentinel: automated device-type identification for security enforcement in IoT (2016). arXiv:1611.04880
40. A. Al-Fuqaha, M. Guizani, M. Mohammadi, M. Aledhari, M. Ayyash, Internet of things: a survey on enabling technologies, protocols, applications. IEEE Commun. Surv. & Tutor. **17**(4), 2347–2376 (2015). http://dx.doi.org/10.1109/comst.2015.2444095
41. D.M. Mendez, I. Papapanagiotou, B. Yang, Internet of things: survey on security and privacy (2017). arXiv:1707.01879.pdf

Chapter 6
Delegation of Access Rights

Abstract This chapter presents a discussion of the critical issue of transferring access rights from one entity to another in an IoT context. The process of transferring such access rights is commonly known as delegation. It is a significant component of an access control system. There have been considerable works in the area of the delegation but much of it assumes a centralized, well-resourced, system and these solutions have limited applicability in the context of the IoT. Where delegation models for the IoT have been proposed they typically provide only coarse-grained control over the delegation of rights. Moreover, many of them require a centralized trusted authority, which can suffer from being a single point of failure. We demonstrate that the IoT requires a secure, flexible, and fine-grained delegation model.

6.1 Introduction

In IoT, amongst other areas, access right delegation is one of the major concerns when addressing security issues [1]. Recall, access rights, in specific, govern who or what can view or use resources by the allocation of rights based on the policies specified. Initially rights are bestowed by the owner of a resource. Delegation allows entities to transfer those rights to other users based on certain policies. In Sect. 2.5 of Chap. 2, we showed how an active entity (e.g. user, device, etc.) can transfer and grant some of its permissions (i.e. access rights) to one or more entities for accessing one or more resources [2]. In large-scale and dynamic systems, e.g. the IoT, delegation is crucial in ensuring flexible, fine-grained and responsive access to resources by allowing users to propagate access in a controlled fashion [3]. The resource-constrained nature of the IoT (e.g. limited battery power, memory capacity, processing power, etc.) and other characteristics (e.g. dynamic nature of the system, mobility in interactions, etc.) make it difficult to employ traditional centralized delegation models. With the exponentially growing number of services, applications and devices, a fundamental issue is to ensure that only the entities that possess the appropriate rights are able to access resources. For the flexible and dynamic management of resources in a collaborative environment, delegation plays an important role in ensuring for the correct distribution of access rights and permissions. Further, the scale and diverse

S. Pal, *Internet of Things and Access Control*, Smart Sensors, Measurement and Instrumentation 37, https://doi.org/10.1007/978-3-030-64998-2_6

nature of the IoT makes it difficult to specify, centrally and in advance, a complete set of access control policies.

Examples of delegation scenarios for the IoT and/or capability-based systems that have been presented in the literature include a friend being given rights so that they can carry out housekeeping tasks while the house owner is away, a mechanic being granted rights to a car's systems to be able to carry out maintenance as directed by the car's owner and the delegation of access to medical records in the event of a patient emergency [4, 5]. As pointed out by [4], access control for the IoT needs to meet the *"dynamicity present in everyday life"*. The nature of the IoT requires a fine-grained approach to access control, including in the handling of delegation. This means, just as with access control itself, the delegation of access rights needs to be governed by policies. We reiterate that the scale and nature of the IoT means that solutions which centralize the control of delegation while depending upon concrete unique identity of the entities are likely to be impractical. Therefore, there is a need for a delegation model that overcomes the limitations of a centralized system and at the same time considers the dynamic characteristics of the IoT.

6.1.1 Problem Description

Previously in this book, we have argued that in a highly dynamic system like the IoT, the interactions between the *things* may be for a very short period of time and may happen only once. It cannot always be known in advance which users will wish to access which services or resources, or which devices will be available at the time access is requested. In other words, the unique identity of the entities that are interacting with one another may not be known in advance, which creates challenges for secure distribution of access rights between those entities. Furthermore, the characteristics of an IoT system e.g. scalability, heterogeneity, mobility, etc. brings new challenges in ensuring the basics of access control, i.e. that only the entities that possess appropriate rights and privileges are able to access desired resources.

Entities still need some basis on which to determine whether to interact, including the bestowal and acknowledgement of access rights. Consider the nature of delegation within the context of the size and scale of the IoT. First, it should not be assumed that entities are constantly in communication with each or the wider system. Secondly, delegation should not depend upon a single central manager. A single control point would run the risks of single point of failure, amongst others. Lastly, in line with other proposals of this book, it should not be assumed that entities making a delegation know the concrete identity of the entities they wish to delegate to.

This last is probably the most unusual assumption that we are making. However, let us consider a non-IoT example. The chair of a committee has to take urgent leave and wishes to delegate their rights to an acting chair. However, the acting chair is not appointed before the chair departs. If the delegation system depended on concrete identity, a delegation could not be made. However, if the delegation system allowed the recipient of a delegation to be defined by attributes, then a delegation could be made on the basis that recipient held the attribute 'acting_chair'. Now consider an example from the IoT. A home owner going on holiday may wish to delegate to their friends control of the automated watering system for their garden. Rather than specifying each friend by concrete identity, delegating control to anyone who holds the appropriate 'friend' attribute would be a more flexible and convenient solution. Note also that in both cases it would be preferable if the entity making the delegation and the entity receiving the delegation did not have to both be simultaneously connected to the system.

From the above discussion we can derive three characteristics that we see as being central to delegation in the IoT.

- *Asynchronous:* The approach to delegation should not require the delegating and receiving entities to be simultaneously active and connected to the system. The delegating entity should be able to specify the delegation and the receiving entity should be able to access the delegation at an arbitrary later time.
- *Distributed:* The system should be distributed and not depend on a single, trusted central point. Any such system would not be scalable to the level required of the IoT.
- *Attribute-Based Identity:* The system should allow entities making delegations to specify the recipients of delegations by listing the values of attributes that the recipients must hold. This allows for more flexibility in specifying delegation.

The first characteristic allows delegation in the IoT to not be bounded by the simultaneous online presence of two entities. In other words, at the time a delegation is made the delegatee may be offline and at the time a delegation is accepted the delegator may be offline. This is practical and reasonable given the dynamic base of the IoT systems where two entities may be interacting for a short period of time and significantly without any previous communications.

As we discussed above, in an IoT system, interactions between the entities may be fleeting, and may occur without any prior knowledge. We contend that centralized architectures are, in general, unsuitable for the IoT as they do not scale to handle the number of devices, users and their associated applications and services to a fine-grained level. This is due to the resource constrained nature of the IoT devices that lack a high volume of processing and storage capacities. At the same time, as discussed earlier, the centralized systems also represent a single point of failure, are heavy-weight in nature and do not necessarily handle heterogeneity in an efficient way [6]. We note that a decentralized approach to IoT delegation can solve several issues discussed above. It has the potential to reduce the costs associated with installing and maintaining such highly centralized server based applications. Instead, we contend that the delegation in the IoT must follow the decentralized nature of the system. This approach will help to achieve interoperability and at the same time will be able to address the scale of an IoT system.

The third characteristic implies that the entities participating in a delegation may not know one another before an interaction occurs. However, an entity wishing to delegate must be able to define characteristics (i.e. *attributes*) of the entity receiving the delegation. In other words, the delegation should be identity-less in nature, which argues for the use of non-unique identities in the delegation. Such a delegation will be more practicable in the context of the extremely large-scale and heterogeneous nature of the IoT than approaches which depend upon concrete identity. In a highly dynamic and large-scale system, like the IoT, it is impractical to know in advance the identities of interacting entities. Note that delegation to particular concrete identities is still possible, through the appropriate choice of attributes specifying the delegation.

Let us now consider some further examples. Recall, the entity that transfers the access right is called the *delegator* and the receiving entity is known as the *delegatee*. Now consider the following two situations.

In the first case, assume, Bob wants a clogged plumbing system in his home to be fixed. He asks for a plumber to come over and inspect the situation. Since Bob works during the day, he needs to setup access for the plumber. This access must be restricted to specific resources (e.g. the front door) and under certain condition (e.g. during working hours). At the same time, he wants to restrict the access to sensitive equipment (e.g. controlling a fire alarm). One solution is to create a single "guest-account" with a predetermined right (and condition) and ask each and every guest, including the plumber, to use this account to access the specific resources in the home. This is however not fine-grained (as it is likely, in this case, to include far more permissions than the plumber requires) and cannot cope with the increasing scale of devices, users and applications as the IoT facilities in the home evolve. Another option to have a predefined set of 'guest categories' but again tailoring them, in advance, to meet the precise requirements is difficult and such situations may not occur often enough to warrant the creation and maintenance of multiple guest accounts.[1] In this situation, Bob can delegate some of his access rights to John (e.g. a plumber) electronically. The right transfer is done dynamically and from within the context of the current rights of Bob. It can be achieved, at a granular level, to transfer exactly the required rights to John. John can in turn delegate some of these delegated rights to some other users if necessary. This process must be conducted in a flexible and secure manner which can be achieved through the following idea *access right delegation must be monotonic, i.e. the delegatee shall have at most the rights of the delegator*.

In the second case, we consider an IoT-based large-scale e-commerce model. Assume, Alice (delegator in this case) has some access right to print a certain number of pages from a dedicated printer. Alice wants to sell her access right to someone who needs the same facility. Suppose Ron (delegatee in this case) is interested in acquiring such an access right for his own printing. However, the entities (i.e. Alice and Ron in this case) do not know one another and do not have any previous interactions. Allowing this is desirable in an extensive system like the IoT where the entities do not have any prior knowledge of one another.

[1]Note, creating and maintaining accounts for every user to every access for all sort of scenarios (e.g. a guest account for regular guests, house keeper, maintenance contractors, delivery services, unexpected visitors, etc.) is a difficult and complex task. As new IoT devices are introduced into the system, access from these accounts need to be maintained individually which greatly impede scalability.

To date, most models for access right delegation in IoT systems are built on the commonly used access control mechanisms e.g. RBAC, ABAC and CapBAC. In Sect. 3.2.2 of Chap. 3, we discussed that these mechanisms have their own advantages and disadvantages when applied to IoT systems. For instance, RBAC provides fine-grained access control over the resources using explicit user-to-role mappings, however, RBAC itself is highly centralized and so is any associated delegation mechanism. ABAC improves policy management by using attributes (e.g. name, age, location, etc.) but at the same time increases the cost of operation when deploying it for a highly scalable system like IoT. CapBAC provides flexible access control. Where a user with an appropriate capability is allowed to perform certain activities over a resource. However, many of the existing CapBAC mechanisms are centralized when validating access rights of subjects. To achieve a more controlled approach to access control, distributed CapBAC models are proposed (e.g. 'DCapBAC' [7]). In this model, validation is performed inside the resource-constrained IoT devices without being any contact to the centralized authority. However, this is in particular challenging for the resource-constrained nature of these IoT devices. In addition, these devices are not always fully trusted and can easily be compromised by the attackers. Delegation in CapBAC systems typically provides for little or no control by the initial owner of the resource. The proposal in [7] does not include an explicit trust model. This means that there is only implicit trust in the distribution and use of capabilities and delegations. There is no means by which entities that are unfamiliar with each other can securely track the progress of delegation through chains of entities. We observe that the establishment of trust in the IoT is challenging as the resource-constrained IoT devices cannot store a large history of transactions. Other issues in building trust mechanisms for such systems include the lack of central trusted authority and the potentially fleeting nature of interactions. Note, in the next chapter (i.e. Chap. 7) we present a detailed discussion on the notion of trust for IoT access control.

We noted in the previous chapters that the scale and diverse nature of the IoT makes it difficult to specify, centrally and in advance, a complete set of access control policies. While delegation allows one entity to grant some of their rights and privileges to another entity, who can then carry out functions allowed by those rights and privileges, it is difficult to establish trust in an IoT environment on a one-to-one basis. Further, the scale of the IoT means that subjects and targets of access control policies are best specified using attributes rather than identities (cf. Chap. 5). This can be extended to delegation, where the possessor of an access right can specify the entities who can receive the delegation by their attributes, not their concrete identity. The IoT requires a fine-grained approach to delegation, including in the handling of delegation in a decentralized environment. On the one hand, this means, just as with access control itself, the delegation of access rights needs to be governed by policies to achieve a distributed and trustworthy access control. On the other hand, the scale and characteristics of the IoT means that centralizing the control of delegation is likely to be impractical. Some additional drawbacks for many centralized systems are high cost of establishment and other privacy issues [8].

6.1.2 Contributions

To overcome the aforementioned issues of transferring access rights, in this chapter, we propose a decentralized, trust-less and fine-grained delegation architecture for IoT using *blockchain* [9]. Blockchain has been seen to have the potential to play a major role in managing, controlling and securing IoT devices with decentralized control, data transparency and auditability. It is tamper-proof where data cannot be manipulated by a malicious actor.

Our primary intention is to facilitate managing and accessing IoT resources without the need for a trusted authority. In particular, we propose and analyze a novel capability-based delegation model for large-scale IoT systems using blockchain. We will show that a blockchain platform provides all the necessary support to implement a delegation process that is identity-less, asynchronous and decentralized by nature. The basic idea is to use capabilities to propagate access rights and use blockchain for communication between various entities. Our design employs attributes for fine-grained access policies and capabilities for access rights transfer. The capabilities are provided to users on request, based on possessed attributes. The delegated capabilities are checked by the IoT *things* upon access. To the best of our knowledge, our proposal is the first one to use capability-based access for the delegation of access rights in IoT using blockchain without relying upon concrete identity. The motivation of our research is also driven by the need for efficient privilege management and policy enforcement in delegation in a large-scale IoT system spanning multiple jurisdictions. Previous work, discussed in the next section (i.e. Sect. 6.2), has not addressed the question of how the potentially vast set of policies in an IoT system (considering the number of devices and the number of users that may wish to access each device) can be managed. We argue that specifying these policies on an individual user/device basis, especially if a priori user identification is required, is not sustainable. Further, we intend to employ attributes for representing an entity and do not consider concrete identity of that entity. The major contributions of this chapter can be summarized as follows:

- We propose a flexible decentralized delegation model for transferring access rights in IoT using blockchain.
- We use blockchain events as capabilities to facilitate access control delegation to IoT devices whereby the generated capabilities are issued by the smart contracts without the involvement of any trusted third-party authentication.
- We provide a detailed description of the system including the architectural design, different interaction patterns between the different entities and the smart contracts.
- Our design is based on attributes and does not depend upon a concrete identity of the IoT *things*, this provides significant flexibility when managing resources at scale.
- Our proposal takes advantage of the decentralized nature of blockchain networks and is asynchronous in nature.

- We demonstrate a detailed proof of concept prototype implementation with Ethereum private blockchain and provide an experimental study of the performance of the proposed model.

6.2 Background

In Sect. 2.2.3 of Chap. 2, we introduced blockchain technology and discussed the emergence of blockchain in IoT access control. In this section, we first discuss the core concepts of delegation related to this research (Sect. 6.2.1). Then we briefly recap the fundamental of blockchain technology (Sect. 6.2.2). Finally, we present available proposals to IoT delegation using blockchain (Sect. 6.2.3).

6.2.1 Core Concepts

Before going to a detail analysis to the blockchain based delegation, in this section, we discuss the essence of delegation in general. We provide the core concepts of delegation and give an example of how it works based to our proposed access control architecture presented in Chap. 3. This will help us to understand the basics of delegation that we suggest in this chapter.

Capability-Based Access: In this book, we used capability for access control and in this chapter, we employ capability for transferring of access rights from one entity to another. Recall, in CapBAC, a capability is the building block of the system's security. It is conceptually similar to a physical key that is used to unlock a door and access the house. There are various significances for the use of capability. We note that, a capability contains four explicit properties: *communicable*, *verifiable*, *secure* and *unforgeable*[2] [4].

- *Communicable:* Since the right has to propagate from the delegators, to the delegates and then to the resources, it should be communicable.
- *Verifiable:* The system should contain enough information about the rights and associated conditions.
- *Secure:* The system must ensure authentication, i.e. capabilities can only be used by the intended user to access the specified resources.
- *Unforgeable:* This indicates that only known and trusted (group of) authorities can produce genuine capabilities.

Whenever we refer to a *capability*, we always assume (or prove) that it possesses at least these four distinct properties. In Sect. 6.4.1, we present a discussion on how these properties can be achieved with our proposal.

[2]Note that these are only the principal properties. In practice, we also require a capability to be *ephemeral* and *revocable*.

From Attributes to a Capability: Recall, access policies denote the permission rules that permit or deny user access to specific resources under certain conditions. The policies and conditions are written by policy administrators referencing various *attributes* about the users, resources, services and the environment. This means that access rights are associated to attribute sets rather than identities (cf. Chap. 5). The point is attributes can be shared while identities are unique to each user. In Sect. 3.4 of Chap. 3, we demonstrated how these attributes are used to generate capabilities for the users.

Delegation Process: Now, we illustrate a simple process of delegation. For simplicity, we recap the example of light-weight key based approach (i.e. symmetric key based approach) that we discussed in Sect. 3.6 of Chap. 3. For instance, a user who wants to delegate a right, needs to generate a new capability from a capability providing that right. The generated capability must adhere to the four properties of a capability we outlined above (i.e. unforgeable, communicable, secure and verifiable). The delegation process can be seen as follows:

$$\texttt{delegated\_capability} = \{K1, \texttt{capability}_1\}$$

where,

$$\texttt{capability}_1 = \{K1, \texttt{delegatedcap}\}_{KS}$$

and `delegatedcap` contains the extra conditions that the capability owner imposes on the delegated right. This is the essence of the *Delegation Rule* that we described in Chap. 3. Note that it also contains the conditions specified by the initial creator of the capability, including any conditions that only apply to the delegated capabilities. The new session key `K1` is generated by the delegator, shared with the delegatee and used as the authentication mechanism associated to the `delegated_capability`. `KS` represents a session key.

JSON Implements Capability: Recall, in Sect. 3.4.7 of Chap. 3, we show how a capability is generated using JSON. In this chapter, we use similar JOSN string for our capability. We use the templates in Fig. 6.1 to show an example on how to encode capabilities and encrypted messages.

These templates show that a message (which can be a request or a capability) has a very simple structure which captures the full expressiveness of other works that use overly complex capability structure e.g. [10, 11]. This structure can be exploited

```
// request, capability or delegation template
{   "attr": <attributes>
    "cond": <condition>  }

// encrypted message template
{   "data": <encrypted request, capability or delegation>
    "capability": <instance of encrypted capability template>  }
```

Fig. 6.1 Simple message templates

```
// Access request without delegation
{    "data":{"attr":{"src":"A","trg":"D","act":"unlock"}}_KS,
     "capability":{
                 "data":{
                   "attr":{"key":KS},
                   "cond":[[src=A],[trg=D],[act=unlock,act=lock],
                   [TIME<01-06-2019 00:00:00]]
                 }_KTH
          }
}
```

Fig. 6.2 An example of an encrypted message without delegation

differently depending upon the kind of information to be encoded. A capability uses the field `cond` to implement the capability conditions. This can capture broad features ranging from time of validity to extra delegation conditions.

In Figs. 6.2 and 6.3, we illustrate two examples of access requests based on the conditions depicted in *Condition Rule* discussed in Sect. 3.4.6 of Chap. 3.

Figure 6.2 does not contain a delegation request and Fig. 6.3 contains one delegation request. The extra condition `[DEPTH<2]` in Fig. 6.3 implies that the delegated right cannot be delegated furthermore. This further emphasises the importance of *Delegation Rule* employed in our access control model. Note, for both of the figures, `KS` represents a session key, `KTH` represents the key belongs to an IoT *thing* and `K1` represents a newly generated session key.

```
// Access request with delegation
{    "data":{"attr":{"src1":"P","trg":"D","act":"unlock"}}_K1,
     "capability":{
               data:{
               "attr":{"src":U,"key":K1},
               "cond":[[src1=P],[act=unlock],[DEPTH<2]]
               }_KS,
               "capability":{
                      "data":{
                       "attr":{"key":KS},
                       "cond":[[src=A],[trg=D],[act=unlock,act=lock],
                       [TIME<01-06-2019 00:00:00]]
                      }_KTH
               }
          }
}
```

Fig. 6.3 An example of an encrypted message with delegation

6.2.2 Blockchain Technology

Blockchain is a tamper-evident, shared and distributed digital ledger of transactions for crypto-currency systems (e.g. Bitcoin [12]). The ledger stores the records permanently in a sequential chain of cryptographic hash-linked blocks. In Fig. 6.4 we illustrate a simple view of blockchain. Every block contains a hash of the previous block. The first block in a blockchain is known as the *genesis* block. It can be seen as block number 'zero' and it does not hold any reference (e.g. a hash value) of a previous block. Instead of having a central ledger with data of the whole system, in a blockchain, every block contains all the necessary data. This enhances the concept of distributed ledger rather than simply creating a centralized one. These blocks are organized by logical time stamps and synchronized among other member nodes within the network.

In blockchain, blocks have a set of transactions (i.e. an agreement or contract) that can be denoted as a small unit of task that is stored in public records. A transaction can be seen as the transfer of values (as price, asset and ownership) between different entities that is broadcast to the network and collected into blocks. These transactions are visible to every node in the blockchain. In other words, transactions are simply the public records of all the executions of data that have ever been performed on blockchain. By design, blockchain does not require any third-party entity to conduct and validate transactions. Everyone participating in the blockchain can validate and verify transactions [13].

Blockchain draws its security from the immutability of the public ledger whose contents are decided via a consensus protocol [14]. Consensus is a fault-tolerant mechanism that ensures that each block in the blockchain agrees on the same state at the time of validating a transaction within it in a dynamic way. The fundamental requirement to achieve a consensus is the uniform acceptance among the nodes on

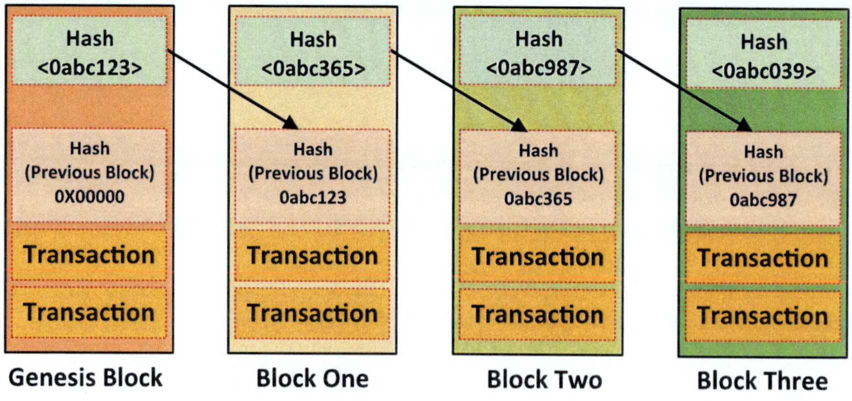

Fig. 6.4 A simple view of formation of blocks in the blockchain

the blockchain of a single data value. Thus, it ensures secure transactions between users in a trust-less network environment.

A blockchain can do more than store blocks of transactions. It can also contain smart contracts which are code blocks that can be executed 'on' the blockchain. This opens new possibilities as blockchain then provides a distributed storage and computational framework on which arbitrary programs can be executed. Some of the important properties of blockchain are summarized as follows [15]:

- *No Central Authority:* Blockchain is able to handle the fast growing number of devices and systems (and their operations) in a distributed and decentralized manner. This will assist in addressing the scalability issues around the IoT. It also eliminates the need for a central, trusted, third-party authority.
- *Consensus:* In a blockchain, for every transaction to be accepted and recorded properly in the ledger, every participant must agree. This is known as consensus. This allows participants to trust the contents of the ledger.
- *Immutable, Irreversible and Tamper-Proof:* These properties of the blockchain ensure that after a transaction is recorded on the ledger, no participant in the network can modify it, as any changes made to the contents of the block will change the hash of the block. It is therefore considered permanent.
- *Accessibility:* In a blockchain, any entity that is a part of the network can read or view any data recorded in the ledger. This enhances the accessibility of data at any time, any place.
- *Auditability:* In a blockchain, auditability of a transaction is provided by the recording of the entity that generated a block. This in turn allows each user in the network to confirm the identity of the source of an entry in the ledger.

Fundamentally, blockchain technology offers a secure and safe way to record and track a list of transactions for large number of devices in a highly transparent, auditable and efficient way by maintaining a peer-to-peer network. This is beneficial when overcoming the limitations of a centralized system for storing information. Transactions in a blockchain are relayed over the P2P network and peers (called as the *miners*) collect these transactions into a data structure—the blocks. The mining technique is called as 'Proof-of-Work' (PoW) which ensures that all blocks miners produce contain valid transactions. As noted above, in blockchain, important components are the data and the hash. Data is collected inside the blocks and the hash is the function that converts an input string of any length (e.g. number or letters) to an output of a fixed length by a hashing encryption algorithm. This improves the security and auditability and in conjunction with the PoW algorithm makes the blockchain immutable. Each block must contain a PoW to be considered valid in the blockchain. Every minor verifies the PoW every time they receive a block. Note, for every transaction the miners are paid a transaction fee [16].

6.2.3 Delegation in IoT Using Blockchain

Several proposals address the specific needs of IoT access control by employing a blockchain network [17, 18]. For instance, Novo [19] presents an access control architecture for scalable management of IoT devices. The architecture is fully distributed in nature leverage the properties of blockchain technology. The access control policies are enforced within the blockchain. The proposed design operates a single smart contract which reduces the communication overhead among the nodes. It also provides access control in real-time to the edge IoT devices. Note, all entities in the system are the part of the blockchain network except the IoT devices. This is due to the resource constrained nature of the IoT devices, where the devices cannot store the heavy-weight blockchain information. The proposed architecture is able to manage a vast amount of IoT devices and provide a decentralized feature of access control that connects a high number of geographically distributed sensor networks. The access control policies are enforced based blockchain technology overcoming the bottleneck of a single centralized authority that manages the access control decisions. In this model, the edge IoT devices do not belong to the blockchain network, they are connected to the blockchain using one or more management hubs. These hubs are distributed over the entire blockchain network and potentially connected in different ways to the IoT devices which significantly provide a considerable flexibility in the overall access control management. Further, this model considers the resource-constrained nature of IoT devices and provide an easier way of integration of the current IoT devices to adapt to the proposed system. However, it does not address delegation at a fine-grained level.

Xu et al. [20] present an access control model for IoT using capability-based access to the resources, called 'BlendCAC'. Based on a blockchain network, a capability-based delegation mechanism is suggested for propagation of access control permissions, where the authorization mechanism of delegation is computed inside the blockchain. The focus on this framework is to build a decentralized framework that overcomes the weaknesses in today's access control mechanisms for the IoT, which are highly dependent upon a centralized authorization server. However, this model's approach to delegation is based on concrete identity, as the recipient of the delegation must be specified when the delegation is created. In [21], the authors further enhance the model discussed in [20] to build an efficient access control system for space situation awareness (SSA) using blockchain. Once again, this approach to delegation is based on the unique identity of an entity.

Manzoor et al. [22] propose a blockchain based proxy re-encryption scheme for secure IoT data sharing in the cloud. The system stores data in a distributed cloud after encryption and to share collected data, the system establishes a run-time dynamic smart contract between the IoT devices and users. This eliminates the involvement of a trusted third party. However, while it refers to delegation it is in the sense of information sharing, not access right sharing and again employs concrete identities. Almoadhoun et al. [23] present a user authentication scheme of IoT devices using blockchain-enabled fog nodes. It improves on the proposal of [22], where fog nodes are used to provide more scalability to the system (by deploying localized computing and processing facilities into the edge of the network) given the resource-constrained nature of the IoT devices. This scheme uses a hybrid architecture of combining blockchain networks involving users, IoT devices and fog nodes. However, these approaches do not discuss the propagation of access right delegation at scale.

Zhang et al. [24] present a smart contract based access control framework for IoT using blockchain. It includes multiple 'access control contracts' for access control between users and resources. In addition, one 'judge contract' is employed to monitor any user for misbehaviour. Finally, one 'register contract' is deployed for controlling the management of both access control and judge contracts. Similarly to our approach, this framework employs blockchain for distributed and trust-less access control for IoT, but unlike our approach, it does not consider the delegation issue at a fine-grained level nor does it envision how to handle identity at scale for delegation. In addition, similar to [20], it does not consider the resource-constrained nature of IoT devices.

With a similar view of [24], Ouaddah et al. [25] present a blockchain-enabled 'token-based' distributed access control framework for the IoT, called 'FairAccess'. In this framework, blockchain is used as a decentralized access control manager. Access control tokens are used for transferring of access rights from one peer to another through a transaction. Significantly, during the transfer of an access token, the appropriate access rights policies are embedded by the users inside the token in the 'locking scripts' (a snippet of code that specifies the type of authorization required for a future transaction) of the transaction output. Thus, when a peer receives the token it must unlock the locking scripts and process the token. However, it uses a public key cryptography-based system for token authorization, which is not an ideal basis for use with constrained IoT devices. Further, the computing capability of the locking scripts are significantly limited.

Tapas et al. [26] discuss a blockchain-based model for authorization and delegation for IoT-cloud services. In this model, authorization is performed inside the cloud and the access control delegation is transferred in terms of a capability. This model enables users to audit the authorization operations without completely trusting the cloud. Similar to [24], this approach does not scale well and once again, entities are explicitly defined by their concrete identities.

Maesa et al. [27] present an approach to creating, managing and enforcing access control policies by exploiting blockchain. In this approach, policies are published on the blockchain and the access right is transferred from one entity to another through a blockchain transaction. The access control policies and the associated access rights exchange are publicly visible over the blockchain networks which provide a transparency where any user can see the policy paired with a resource and the subjects who currently holds the rights to access the resource but this does pose privacy issues. Unlike [26], it does not use capabilities for access right delegation and depends upon a highly centralized XACML reference architecture.

Ali et al. [28] discuss an approach of permission-based delegation and access control for blockchain-based IoT systems. The proposed approach leverages the decentralized nature of blockchain for permission delegation between the entities within the system. The main motivation of this study is to establish trust between the entities while removing the central trusted authority to maintain the trust degrees for each entity. This is achieved by the use of PoW—the consensus mechanism of blockchain. To provide a light-weight solution, access right delegation is assigned to a node with a minimum number of permissions. No proof of concept implementation is detailed.

Nuss et al. [29] present a blockchain based identity and access management framework for large-scale IoT systems. This proposal first investigates the current challenges of identity and access management issues in IoT and then employs blockchain to examine how those challenges can be controlled. The proposal addresses the increased demand for secure and comprehensive identity and access management issues in IoT as well as discuss the question of interoperability between heterogeneous and resource constrained devices using blockchain technology. Further, it addresses the scalability issue in terms of network and storage consumption. However, the actual design and implementation is not provided.

Shafagh et al. [30] present a blockchain-based design for the IoT systems that brings a distributed access control and data management. The authors identify three requirements that are essential for such design, they are: secure data storage, IoT compatibility and decentralized access control management. The design enhances blockchain technology to manage ownerships and data sharing between the owners and the IoT devices. Owners can create new transactions to the blockchain that contains the identifier of the data stream and the service's public key. When a user wants to revoke an access right from a specific user, it changes the encryption key and shares the new key with all authorized services over the blockchain network, except the one is revoked. Once again, no implementation detail of the design is provided.

Le and Mutka [31] present an IoT access control mechanism, named 'CapChain'. This allows users to share and delegate their access rights efficiently and seamlessly to IoT devices in public but still maintain privacy and user's identity by secure distribution of keys leveraging the use of blockchain transactions. It uses capabilities for access right delegation over the blockchain networks. Here, every IoT device in the network contains at least one owner who has a full control over the device and is capable of generating capabilities based on the access control policies specified by the system. The capabilities are then transferred to one device to another via transactions. An experimental setup is provided with evaluation results to support the design.

While there have been several works that discuss access control issues in IoT using blockchain networks, we observe that the specific issue of delegation is overlooked. While a few initial approaches discuss the issue of delegation at a primitive level, they do not consider the specific characteristics of an IoT system, nor do they provide any implementation detail.

6.3 A Motivating Scenario

In this section, we illustrate a practical use-case example of a delegation scenario for transferring of access rights. In Chap. 3, we employed a use-case scenario of smart healthcare. However, to show the robustness of our proposal and IoT application domains, in this chapter, we employ a scenario of an Internet business model where the core of the model are third parties, commonly known as the *brokers*. The brokers bring *buyers* and *sellers* together and perform an online transaction according to the policies. Note, our proposal could easily be incorporated into any typical smart healthcare scenario, or any other situation where funds transfer is not involved (discussed in Sect. 6.7).

Now, let us consider the following case where three parties are involved, namely, *owner* (owner of a resource), *broker* (facilitates access to a resource) and *consumer* (who wants to access a resource). In Fig. 6.5, we depict the use-case. Suppose, the owner has a business that rents on-site printers in a particular university and can provide access rights for printing documents on any of the deployed printers. This business wishes to transfer (by selling) some access rights of printing to a consumer (for instance, a student at the university). The business advertises this information and a broker (e.g. a student association) buys the rights which are then resold based on their business policies. In both cases, the owner and the broker charge fees to at least one party involved in the transaction. In this case, the owner charges to the broker and the broker charges to the consumers. Next, we explore why delegation is a good solution for this scenario.

The owner still retains full access and ownership of the deployed printers and provides access rights to a broker by generating a capability (i.e. an access token). The broker, in turn, will delegate the provided access rights to the consumers (using an access token). This is where two significant properties of our delegation approach (i.e. identity-less and asynchronous communication) and the use of blockchain, are

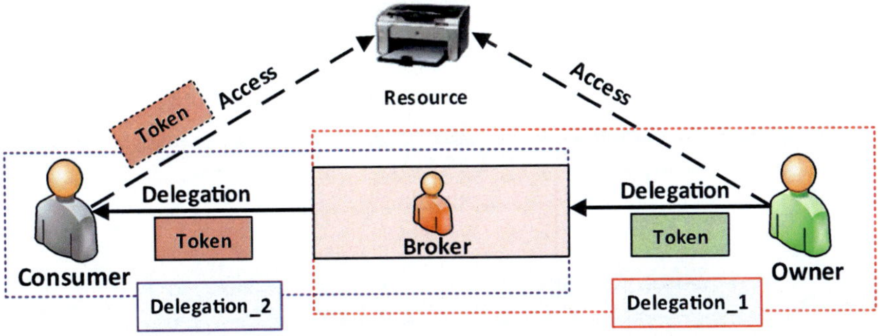

Fig. 6.5 The process of delegating an access right. Owner contacts consumers for selling of access rights. Owner delegates an access right in the form of a token to a broker (Delegation_1) which further issues another token (Delegation_2) to a specific consumer. Now the consumer is able to access a resource (e.g. a printer) with the delegated token issued by the broker

most useful. Firstly, the broker will not know in advance to whom he or she is going to delegate the access to the printer. In a traditional access control framework, the delegation process is usually initiated by the delegator knowing the identity of the delegatee but this approach does not work in our case. The broker cannot know in advance who all its customers are and allowing individual access on request requires synchronous communication and a multiplicity of access policies. Thus, the access control rules must be written in a way that is independent of explicit user identities. In addition, the rules for delegation are stored within the blockchain which means that every party can check and audit the contracts.

In this use-case example avoiding the use of concrete identities by employing attributes could be beneficial as it avoids the need to specify the customer's identity in advance. The access control policies are evaluated based on such characteristics as access time, location, age, student enrolment status, etc. Next, the delegation process must be asynchronous as it is initialized by the delegatee. Precisely, the broker needs only to setup an initial 'delegation service' and the access right will be delegated on demand without any further interactions. This can be achieved if the 'delegation service' is written as a *smart contract* on the blockchain. Finally, this particular scenario ultimately requires the automatic handling of financial transactions which is the primary use of blockchains e.g. Bitcoin, Ethereum, etc. [13]. Using crypto-currencies, the students are able to send appropriate access fees to the broker's smart contract which will in turn delegate access to the printer and forward any required access fee to the printer owner's account. Note, a smart contact can be viewed as a special account consisting of functions (i.e. code) and data (i.e. its state). Thus, a blockchain-based delegation will possess the properties that we propose as well as other important features e.g. the decentralization of the system, and the ability to intrinsically process financial transactions.

6.4 Proposed Delegation Architecture

In this section, we discuss our proposed delegation architecture. First, we discuss the basic properties of a delegation that are supported by the architecture. Second, we discuss the essence of secure right delegation. Third, we present an overview of the architectural components. Finally, we illustrate the communication protocol satisfying an access request.

6.4.1 Delegation Properties

A delegation process must allow the control of two fundamental properties, namely, the *right to delegate* and the *delegation of right*.

The *right to delegate* controls the transfer of rights from the delegator to the delegatee. This includes restrictions e.g. the delegatee's ability to delegate the delegated right up to a certain depth (i.e. the number of delegations). An obvious example in our use-case is that the consumers who bought a right to print should not be allowed to delegate (or resell) that access right any further. In practice, this property is important to verify deep delegation chains as it ensures the termination of the validation of the right transfers.

The *delegation of right* controls the transfer of the rights from one entity to another. In practice, this is achieved through delegation conditions which must satisfy certain monotonicity properties to safeguard from possible attacks e.g. privilege escalation. These monotonicity properties must be verified at delegation or access time and this ensures the correctness (or security) of the delegation process. The delegation of rights is quite distinct from traditional delegation because, in our setting, the delegators do not know in advance to whom they will delegate the access rights they possess. Hence, the best means to achieve this is by imposing certain delegation conditions that are at the same time generic and granular. In addition to a controlled delegation, it is also necessary that a delegation process satisfy the practical and security properties (i.e. *communicable*, *verifiable*, *secure* and *unforgeable*) that we discussed in Sect. 6.2.1.

In a traditional implementation of a token-based access control mechanism, e.g. CapBAC, the capability token possesses these four properties [4]. However, as noted above, CapBAC does not provide other properties that we require (i.e. *ephemeral* and *revocable*). In this chapter, we use blockchain for delegation, which is employed as the medium to generate, communicate and verify these properties. Specifically, access control tokens (delegated or not) are implemented as *events*. In a blockchain network, an *event* is a special form of transaction that is generated and linked to a smart contract. The fact that each event inherently comes from a given smart contract is crucial in implementing an efficient verification of the delegation. An event inherits all the important security properties of a blockchain transaction e.g. *ownership* (i.e. it is owned by the smart contract which generates it), *immutability* (i.e. the deeper the block containing the event is in the blockchain, the harder it is to alter) and *shared* (i.e. it is automatically accessible to all parties involved in the blockchain network).

The ownership and immutability alone ensure that events are also unforgeable and secure. In fact, an event generated by a smart contract will be securely attached to that contract. If the address (public key) of a delegatee is recorded inside of that event and since that event is immutable, then only the delegatee can use that event by proving that he or she owns that public key. An event is obviously communicable since it resides on the blockchain and everyone can access it. Finally, it is also verifiable by simply checking that the event is issued by a 'valid' smart contract and the delegatee address recorded in that event can be trusted as well because it is immutable. In a delegation chain, verification is achieved by traversing the chain in a backward manner.

6.4.2 Secure Right Delegation

Given the uncertainty in interactions present in an IoT system, our intention is to use access delegation to transfer rights without any prior trust establishment between a third-party and the central system or the resources. Intuitively, the system trusts a right holder and some of that trust can be transferred to a third-party who is trusted by the later. The transferred trust implicitly propagates through the system making the delegatee indistinguishable from any other registered user.

Similar to dynamic attributes (discussed in Sect. 3.4.2 of Chap. 3), delegation happens after a capability is generated since a delegator usually does not know in advance to whom they will delegate some right to. More precisely, the decision as to whether a capability can be delegated or not can be inferred during the authorization process but the act of delegating an access right occurs well afterwards. If the delegation is allowed, the act of delegation itself must be initiated by the rights holder and a capability encapsulating the delegated right must be generated for use by the third party. The generation of that capability must follow a critical property of delegation, namely *monotonicity*. This property must be satisfied and verified by the central authority or the resource at access time, depending upon the implementation. In essence, we need to ensure that the delegation process must generate a capability which follows the four properties discussed above, namely *unforgeable*, *communicable*, *secure* and *verifiable* and the generation process itself must adhere to some *monotonic delegation rules*. The capability associated with a delegated right is functionally indistinguishable from any other capability circulating in the system.

A monotonic delegation process ensures that capabilities are generated safely in order to avoid security issues e.g. privilege escalation. This form of delegation has been rigorously formalized in other works. For instance, [32] provides simple logical rules that governs monotonic delegation of roles which are summarized as follows: (a) the preset delegation depth decreases through nested delegation process, (b) weaker roles can be delegated by holders of stronger roles and (c) stronger conditions can be imposed when delegating roles.

Note, one of the major differences between our approach and that of [32] is that they are delegating static roles while we are delegating access rights associated to dynamic capabilities.

6.4.3 Overview of the Architectural Components

In Fig. 6.6, we illustrate the proposed delegation architecture. It consists of the following components:

- Brokers: Users who sell an access right. Brokers can be an individual or a group of users. Brokers can be seen as the *delegator*.
- Buyers: Users who is willing to buy a delegation right from a broker. Buyers can be seen as the *delegatee*.

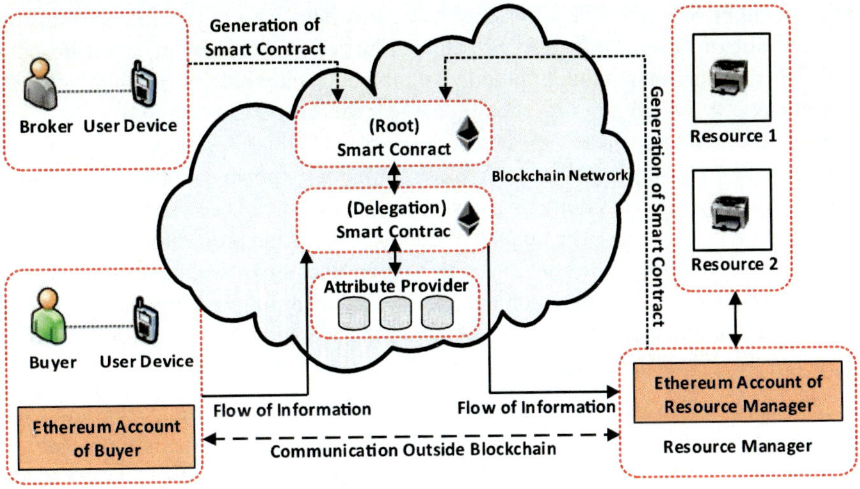

Fig. 6.6 The proposed architecture of blockchain-based delegation process

- User Device: It remains the same that we discussed in Sect. 3.4.4 of Chap. 3. Recall, it is the smart mobile device carried by the human user (e.g. smart phones, laptops, tablets, etc.). These devices can be used to access a resource, deploy smart contracts and communicate between the users.
- Attribute Provider: It is a storage repository. It stores attributes (e.g. student enrolment status, class, age, year, etc.) corresponding to a principal and supplies attributes associated with the specific user addresses for authorization. Since this data will be stored on the blockchain, it can be as simple as a key-value mapping for each attribute class.
- Resource: A device (or a group of devices) that offers specific services, and therefore is able to perform some operations once receive a user's command. They reside outside the blockchain network and connected to a resource manager. In our running example, we do not specifically distinguish between the resource and the resource manager.
- Resource Manager: One or a group of entity responsible for managing access permissions of a set of IoT devices. In other words, it manages one or a group of resources (e.g. printers) where several devices are connected. In general, the resource managers are considered light-weight nodes in our architecture. Note, they do not store the blockchain information or do not verify the blockchain transactions as the *miner* nodes perform. Significantly, they also do not need to be connected to the blockchain network constantly. All the registered IoT devices in the system must belong to at least one registered resource manager, and an IoT device may belong to multiple resource managers at the same time. A resource manager is connected directly with a blockchain network. One resource manager can hold many resources, or many resources can be connected to one resource manager.

- Blockchain Network: We choose a private blockchain for the purpose of testing and evaluating the system. This will allow us to get more consistent, controlled and reliable evaluation results. In practice, the blockchain would be a public network that any peer can join in order to request delegation and access controls.
- Smart Contract: It is a self-executing piece of code that represents an agreement between two or more parties. Unlike the commonly known contracts that depend upon the reputation of counter parties, a smart contract can be created by untrusted and anonymous users. The smart contracts are the source of autonominty in the network as successful transactions will trigger other smart contracts or generate a network wide event from which peers can react accordingly. Note, the delegation architecture that is discussed in this chapter is governed by the operations defined in two smart contracts triggered by the blockchain transactions.

The user device (for broker and buyer) is able to generate and communicate to the (delegation) smart contacts reside in blockchain. The attribute provider is capable of communicating directly to the (delegation) smart contacts. Importantly, the buyers and the resource manager hold their corresponding Ethereum account for transactions. Notably, the communication between the buyers and the resource manager is executed outside the blockchain network.

6.4.4 Communication Protocol

Now we discuss how the various components in the architecture would interact. In Fig. 6.7, we illustrate the protocol for satisfying a user request. It highlights a two-stage communication protocol, where the first stage (steps 1–5) consists of the delegation process itself and the second stage (steps 6–8) captures the access to a resource. Note, in this communication, an entity can be seen as both the delegator and the delegatee. For example, a broker can be a delegator and a delegatee based on the actions it performed. Therefore, to avoid confusion in the representation of the delegation process, we use the following, *resource owner* as the *delegator*, *broker* as the *delegatee* and *customer/end-user* as the *user*. Indeed, a user is a delegatee. Next, we provide a detailed discussion of various communication steps.

- Step 0: We assume that the attribute providers are already deployed into the blockchain network in the form of smart contracts acting as attribute storages. In their simplest form, they contain key-value pairs of account addresses and attribute values.[3]
- Step 1: The delegator deploys a smart contract which will become the root of all delegations of rights to access a resource. This smart contract stores the 'trusted' addresses of attribute providers and accepts transactions for access or delegation.

[3]Note, there is no privacy whatsoever regarding these attributes since we assume that they are publicly available. Thus, anyone can cross-compile all attributes associated to a given address. There are ways around this using more involved blockchain constructs but this is specific to the blockchain implementation itself.

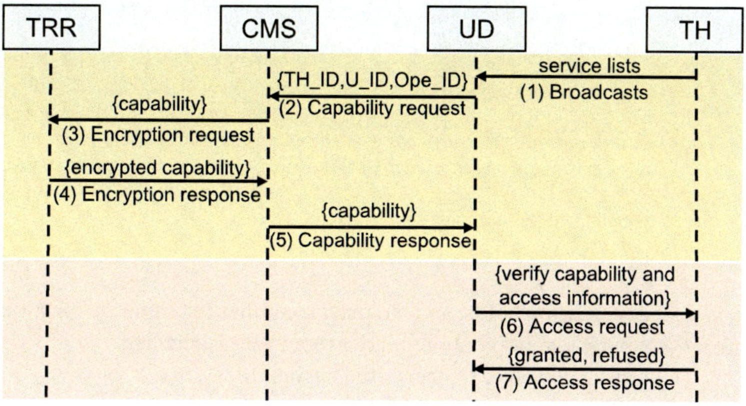

Fig. 6.7 The communication associated with an access request. The horizontal dotted lines denote accessing or generating an event

- Step 2: The delegatee deploys a delegation smart contract with the address of the root smart contract and requests a delegation right by sending a transaction to the delegator's smart contract. If granted, the smart contract of the delegator generates an event (Step 2a). This 'proves' that the delegatee has access to a resource under certain conditions. This condition reflects the right that the delegator holds when accessing or delegating access to the resource.
- Step 3: User wants to access a resource, and therefore the user requests access via delegation from the smart contract of the delegatee by triggering a transaction.
- Step 4: The delegation smart contract issued by the delegatee in Step 2, requests attributes held by the user from a trusted attribute provider (Step 4a), by executing another smart contract. This later smart contract is one of the attribute provider smart contracts available within the blockchain network as discussed in Step 0. The attribute provider smart contract returns the requested attributes to the delegation smart contract of the delegatee (Step 4b).
- Step 5: The delegation smart contract of the delegatee executes logic that checks the fetched user attribute values. If all the required attributes are valid and the (static) delegation condition holds then a new event is generated by the delegatee smart contract. This event signifies that the user can employ the delegated rights.
- Step 6: At this point, the user should be aware whether his/her access request has been granted or denied by observing the generation of the above event. If successful, the user finally requests an access directly to the resource. This request contains the address of the delegatee's smart contract.
- Step 7: The resource verifies that the public key used to sign the request belongs to the requesting user by ensuring that the latter possesses the private key associated with that public key. This can be implemented as a simple cryptographic challenge response (Steps 7a and 7b).
- Step 8: The resource scans through the events generated by the delegatee's smart contract. It then finds the most recent one reflecting the user's address, verifies

```
event printer_access_request(
      address requester (0x01),
      address parent (0x00),
      bytes conditions (<cond>)
  );_0x00
This event is emitted through the smart contract code:
emit printer_access_request(requester,get_authority(),conditions);
```

Fig. 6.8 An event generated by the smart contract 'R'

that the access condition holds, and recursively applies the same verification to the events of any parent smart contract until a root event is reached.
- Step 9: The user is now able to access the desired resources.

Now consider the example of accessing a printer (i.e. a resource in this case) that we discussed in Sect. 6.3. We assume that the attribute providers are already deployed into the blockchain network. Suppose, a resource owner deploys a smart contract 'R' into the blockchain network. In our case, it is the root (smart contract) of all delegations of rights to access the printer. Therefore, R stores the 'trusted' addresses of attribute providers and accepts transactions for access or delegation. Now, a resource broker deploys a delegation smart contract 'D' with address `0x01` and requests a delegation right by sending a transaction to R by generating an event as shown in Fig. 6.8. The conditions (denoted as `<cond>`) in Fig. 6.8 could be a specific *Delegation Rule* similar to that discussed in Sect. 6.2.1.

This piece of code shown in Fig. 6.8 states that the broker has access to certain printers under certain conditions. This condition reflects the right that the broker holds when accessing or delegating access to the printer. In particular, these conditions may implement the right to delegate together with some additional conditions e.g. colour and B&W printing, etc. We call this event (of Fig. 6.8) a *root event*. Root events are always generated by the resource owners.

Now let us consider, a user (e.g. a student), with account address `0x02` wants to use a printer, and therefore requests access via delegation from smart contract D by triggering a transaction (using his/her own mobile device). In this case, the delegation smart contract D requests attributes from a trusted attribute provider, which is another smart contract 'A'. The later smart contract (i.e. A) must have been pre-deployed and available within the blockchain network. One the request arrived to the attribute provider, the attribute provider returns the requested attributes to D. The delegation smart contract D executes logic that are employed to checks the fetched attribute values. If all the required attributes and conditions are valid then a new event is generated by D as shown in Fig. 6.9.

In this situation, the user should be aware whether his/her access request has been granted or denied by observing the generation of the above event (of Fig. 6.9). If successful, the user finally requests an access directly to the printer by submitting an access request containing the address `0x01` of the smart contract D along with the documents to be printed. As noted earlier, the resource (i.e. the printer) verifies that the address (i.e. public key) belongs to the requesting user by ensuring

```
event printer_access_request(
      address requester (0x02),
      address parent (0x00),
      bytes conditions (<cond>)
  );_0x01
This event is emitted through the smart contract code:
emit printer_access_request(requester,get_authority(),conditions);
```

Fig. 6.9 An event generated by the smart contract 'D'

```
def get_trail(req_addr, contract_addr):
 event = get_latest_event(req_addr,contract_addr)
 if event is None:
   raise Exception("No event!")
 attrs = get_attributes(event['args']['recipient'])
 if not satisfy(attrs, event['args']['cond']):
   raise Exception("Condition violated!")
 if contract_addr == w3.toChecksumAddress(address):
   return [event]
 parent = event['args']['parent']
 trail = get_trail(event['address'],parent)
 trail.append(event)
 return trail
```

Fig. 6.10 Function that verifies an access to a resource (e.g. a printer)

that the latter possesses the private key associated with that public key. This can be implemented as a simple cryptographic challenge where the printer encrypts a nonce (`enc(nonce,PKU)`) with the public key (`PKU`) of the user and asks the user to decrypt it. In general, this step involves a two-way authentication. Since this is achieved outside of the blockchain network, a TLS connection can be firstly established between the printer and the connecting user device which validates the identity of the printer. Then, the cryptographic challenge can be conducted to verify the identity of the connecting user device. The resource scans through the events generated by the smart contract D. The resource then finds the most recent one addressed to the user's address, verifies that the access condition holds, and recursively applies the same verification to the events of any parent smart contract until a root event is reached (we provide a sample python code for this process in Fig. 6.10). Finally, the user is able to access the desired printer.

6.5 Implementation

In this section, we provide a detailed description of the implementation for our proposed delegation architecture. To demonstrate the feasibility of the architecture, we have implemented the design using a private Ethereum blockchain. Ethereum is a distributed computing platform and operating system that provides smart contract functionality. Ethereum fundamentally is a public blockchain network, however, it offers a simple and elegant development environment that allows for the testing of applications built on Ethereum. When it is used to create a private testing blockchain network, it can be seen as a decentralized generalized transaction ledger that provides access to a distrusted virtual machine and smart contracts. Our prototype was built on a 2.2 Ghz Core 2 Duo MacBook pro with 16 GB of memory. We have deployed three nodes consisting of one Ether miner and two peers. The network was deployed and managed with Geth v1.8.13 and the smart contracts were compiled with *solc* 0.5.0 (i.e. *Solidity* version 0.5.0). Some of the client scripts are written in python 3.7 where we have used the Web3 package version 4.8.3 to interact with the blockchain.

Solidity is an object oriented programming language that is designed to be executed on the Ethereum Virtual Machine (EVM) [33]. It provides a high level language through which the logic governing smart contract accounts can be written with ease. The solidity program is compiled into binary codes that the EVM executes 'on top' of the blockchain network.[4] Solidity is a Turing complete language, meaning that it can emulate all computable functions. However, executing an arbitrary function on the blockchain can be difficult and expensive so it is important that the smart contract is governed by simple core logics and any extra processing be done outside of the blockchain network. In particular, though theoretically possible, encrypting or decrypting cipher texts in a smart contract is not advised while computing and verifying hashes are already implemented natively and efficiently. The smart contracts should try to be as simple as possible in order to minimize cost to users and resource owners. Smart contracts are best utilized as 'smart' storage containers that can perform logic operations on the data it has stored. It can also perform simple verification processes on incoming data before it is stored. As smart contracts can be deployed in a public blockchain network.

The first component in our implementation is the smart contract R which is deployed by the resource owner. It maintains a list of trusted attribute providers that are used to collect attribute values when processing access or delegation requests. When a request arrives, the corresponding attributes are queried from the attribute providers and the access and delegation conditions checked. In Fig. 6.11, we show a simple implementation of such a process using Solidity. The `condition()` function should not contain any computationally expensive segments because, in Ethereum, each atomic operation is paid for in *gas* (it is the execution fee for each operation made and decided by the miners). If the collected attributes satisfy the condition then a new transaction is fired (e.g. the access or delegation fee is forwarded to the resource account) and an event is emitted. This event records the address of the requestor `req` and the condition `cond` of access. The event that is emitted can be registered by off-chain systems and processed accordingly. In the context of a delegation system the event generated by a smart contract giving a user resource access can be found by the resource to verify that access is legitimate. However, in

[4] In reality, the binary code is firstly executed by a single miner which manages to mine the next block in the blockchain. When that block is propagated, each node will also execute that function locally.

```
function access_request(address req) payable public {
  attrs = attr_provider.get_attributes(req);
  if (condition(attrs)) {
    printer.transfer(msg.value);
    emit access_event(req,address(this),cond);
  }
}
```

Fig. 6.11 A smart contract function that is executed when providing access or delegation

theory, these events can be used to generate off chain transactions or trigger other smart contracts. Events are also how most multi-layered blockchain networks handle inter-chain communications.

Recall, we discuss among others, two distinctive properties of our delegation process are, identity-less and asynchronous. That is, the delegator does not know in advance to whom (and when) the acquired rights will be delegated. To achieve this, the delegator (i.e. the broker in this case) creates the smart contract D and requests access by triggering the function `access_request()` (as shown in Fig. 6.11) of the smart contract R. The event emitted by R above will serve as a *proof* that D is indeed valid. The smart contract D exists on the blockchain waiting for consumers (e.g. students) requiring access to a resource. When a resource access is requested through D, it runs a similar function (cf. Fig. 6.11) except that the event emission instruction is replaced with:

```
emit access_event(req,get_authority(),cond);
```

where *req* is the address of the consumer requesting the access. This line of code instructs the smart contract to generate an event where the `parent` field is populated by the address of the 'parent' smart contract that *proves* the validity of D. This address is fetched by the function `get_authority()`. A crucial observation here is that the delegation contract must use a subset of the attribute providers trusted by the parent smart contract. This will ensure the important security properties provided by the monotonic delegation (discussed in Sect. 6.1.1).

Finally, at access time, the consumer must be sure that D has successfully emitted an event recording its address. The consumer submits the address of D and his crypto currency account address to the resource. The resource verifies that the consumer is indeed the owner of that account via an extra cryptographic challenge discussed earlier. If positive, then the resource scans all events emitted by D and finds the most recent one addressed to the user. If this event is found and it is not a root event, then the printer will scan the events of the contract with address `event.parent` for the most recent one reflecting to D, and so on. Note, this is a simple recursive lookup with time complexity $\Theta(n \times k)$ where there are k contracts in the delegation chain and each contract contains on an average n events. In practice, the number of events per contract n is much larger than the delegation depth k so that looking for the most recent event only provides some early optimization. In our prototype, this recursive lookup is implemented in python and executed by the resource or resource manager (as shown in Fig. 6.10 of Sect. 6.4.4).

Note, in our implementation we employed an Ethereum private blockchain network for generating and distributing smart contracts. However, our conceptual framework can be implemented using any blockchain network that supports a sufficient level of smart contract execution. However, the design could not be directly supported by a 'Bitcoin' blockchain platform as Bitcoin does not support smart contract functionality. Note, in our proposal, we assume the use of a private blockchain to ensure a level of protection for the attributes as compared to storing them on a public blockchain. However, a public blockchain could be used if privacy is not seen as a priority or if there is no private information stored in the delegation system at all.

6.6 Evaluation

In this section, we present the evaluation of the proposed implementation. In Fig. 6.12, we illustrate the achieved results. Times are measured in both milli-seconds (ms) and seconds (s) respectively. The median value of time taken for processing steps 1–5 (of Fig. 6.7 discussed in Sect. 6.4.4) of the delegation process is 6 ms with the majority of proof of access events are generated within 3.7 ms of that time. We note that this speed is not entirely surprising given the fact that in our controlled environment, the two transactions producing that event are mined almost immediately. However, in a live blockchain environment, blocks containing new transactions are mined in certain period of time. In a public Ethereum blockchain, the default difficulty of the proof of work is set so that blocks are mined about every 15 s. Thus, in our particular case, user would need to wait at least 30 s for the event to be generated. It is usually more prudent to wait for a few extra blocks to ensure that the transactions events are deep enough within the blockchain which increases the certainty that these new objects do not live on a stale branch.

The verification of access (step 8 of Fig. 6.7) takes 1.25 ms for a direct access (i.e. without delegation) where we assume that there are 500 root events (and the event giving the access is the oldest). This time, however, increases with the number of events and the depth of delegation. In Sect. 6.4.4, we stated that the verification process (of Fig. 6.10) has a run-time proportional to the product of the delegation depth and the average number of events attached to each delegation contract involved in the chain. This statement is supported by our evaluation results illustrated in Fig. 6.12. For instance, with a delegation depth 0, average time taken to generate 50 events is 0.17 ms and for 500 events it is 1.25 ms. Similarly, with a delegation depth 3, average time taken to generate 50 events is 1.65 ms and for 500 events it is 5.61 ms. We observe that the effect of deep delegation on the access time is more pronounced i.e. the slope of the curves become steeper with increased depth of delegation. In practice, the number of events is much larger than the delegation depth but a delegation of depth k will still result in an access time that is k-fold larger

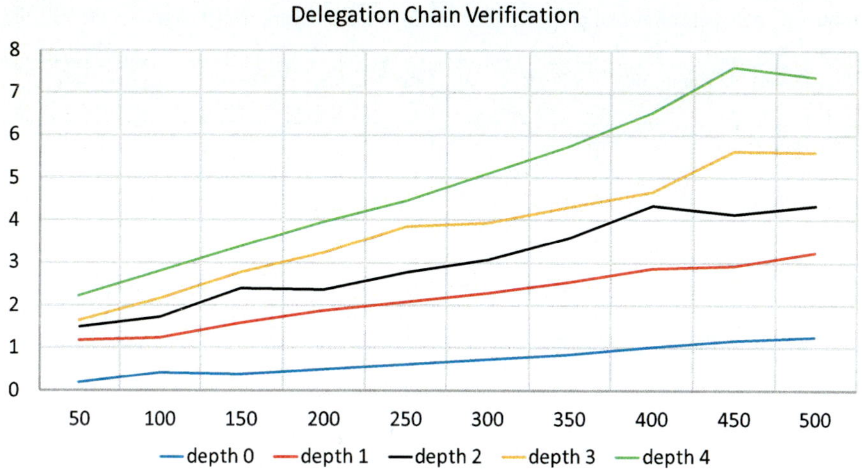

Fig. 6.12 Time taken to verify delegated accesses. The X axis represents the average number of events generated by each delegation contract and the Y axis represents the time spent (in ms) to verify the delegation chain

than an undelegated access to the resource. However, we expect that finding the most recent event only may provide some optimization.

The evaluation results reinforce the feasibility of our design in a practical situation. In particular, we can implement a delegation process satisfying our devised properties which include identity-less, asynchronous and distributed. Note, in this chapter, we do not explicitly discuss the issue of access rights revocation. However, this can be added on the top of our framework by using standard restrictions e.g. temporal access conditions. A more aggressive revocation can also be implemented as a list of revoked delegation contracts which the resource manager keeps track of. Thus, in addition to verifying the delegation chain, the resource can also perform a lookup on the current smart contract addresses to ensure that they are still allowed to delegate an access. This would invalidate all events relying on a revoked smart contract, even if they are not issued by the same smart contract. However, this solution is not fully ideal since the owner of the revoked smart contract may simply create a new delegation as long as the owner still possesses the required attributes. Emphatically, the identity-less property limits the power of revocation as access is not attached to identities any more.

In our design, we do not use the blockchain to provide the access control functionality. We used a fully-featured ABAC system that is able to handle the policy management and identity of an entity to a more flexible and fine-grained level based on attributes. Without the use of the ABAC system, the root smart contract can record access control rules based on the attributes, but the access policy management is very limited. Another limitation is that there is no support for an XACML implementation of ABAC for blockchain. More importantly, we did not want to tie our delegation

Table 6.1 Comparison of the proposed blockchain-based delegation approach with the existing approaches

Properties	Ramos et al. [7]	Xu et al. [20]	Xu et al. [21]	Tapas et al. [26]	[Current]
Non-unique identity					✓
Asynchronous		✓	✓		✓
Decentralized		✓	✓	✓	✓
Monotonic	✓				✓
Auditability		✓	✓	✓	✓
Use of capability	✓	✓	✓	✓	✓

model to a particular method of managing the capabilities. Our approach can be used with both blockchain-based and non blockchain-based access control models.

Our proposal improves upon previously proposed access right delegation approaches for the IoT. In Table 6.1, we provide a brief summary comparison of our proposal with previous work on delegation in the IoT. We compare it both with both blockchain-based and non blockchain-based approaches. We use a few characteristics for comparison e.g. non-unique identity, asynchronous, decentralized, monotonic, auditability and use of capability. Notably our proposal is identity-less and used attributes to authenticate a legitimate user in the system for distribution of access right delegation. As discussed earlier, these are highly beneficial in an IoT context.

6.7 Discussion

Now we provide a general discussion on how our proposal would work in different use-case scenarios apart from the Internet-based business model that we discussed in Sect. 6.3. Let us consider the following use-case. Gusmeroli et al. [4] describe a situation where Bob, a car owner, wishes to have his car serviced by Dave, a mechanic. Access to the car engine data requires a capability which has been issued to Bob by the car manufacturer. Now Bob can create a delegated capability for Dave. In Gusmeroli et al.'s system [4], the only limits on delegation are whether the capability may be delegated and the length of the delegation chain. There is nothing in the system of [4] that the car control unit can do to ensure that Dave is a qualified mechanic. Unlike that system, in our proposal, the *Delegation Rule* can contain a test to ensure that Dave possesses the appropriate attributes or other conditions (discussed in Sect. 6.2.1). Dave passes this attribute, along with the delegated capability, to the car control unit, proving that Dave is authorized to perform the operations required.

Our delegation model can adapt to this as an overseeing mechanic institution could act as the attribute provider and provide the appropriate attributes that proves that

Fig. 6.13 Format of event for giving a mechanic access

```
event car_access_request(
      address requester,
      address owner,
      bytes conditions
      );
```

Dave is qualified. These attributes can be trusted as they come from a validated source in the institution. This can be achieved using a similar event depicted in Fig. 6.13, and the corresponding attributes can be recorded as depicted in Fig. 6.14. Note, in blockchain a capability is processed as an event.

The next use-case is drawn from [5]. In this example Alice is a patient of general practitioner Fritz, who works in clinic C. Fritz determines that Alice needs to be referred to hospital H. At the hospital Alice is examined by George, a doctor working for the hospital. Fritz wants to allow George access to Alice's medical records held by the clinic. George determines that Alice is need of surgery and consults with Hilary, a surgeon employed by the hospital. Alice is admitted and Hilary wishes to give Fritz access to her diagnosis and treatment plans. Delegation is necessary in this case, as the medical records and treatment plans are dynamic and therefore passing copies around would produce a difficult to control version problem. To allow the doctors in the hospital access to Alice's file from the clinic, Fritz creates a delegated capability and passes it to George, who may then pass it to Hilary. Here is where Fritz would create an event on our framework, showing that George and Hilary have access to the files. Similarly, Hilary can create a delegated capability for Alice's diagnosis and treatment files and pass that to Fritz. Hilary creates an event to give Fritz access to her files. In our proposal, Fritz would have included delegation rules in the delegated capability ensuring that Alice's files can only be viewed by doctors from the hospital who are assigned to Alice's case. In Fig. 6.15, we show a simple example outline about who maintains and controls the attributes.

The doctors would prove their valid use of the capability by possession of the appropriate attributes (doctor or surgeon at the hospital and assignment to Alice's case). In the proposal of [5], Fritz simply delegates the capability to George without

```
int[] certificates;
string[] mechanics;
string[] cars-owners;
string[] cars;
mapping(mechanics => certificates) proof-of-mechanic;
```

Fig. 6.14 A simple outline for recording attributes

Fig. 6.15 Format of event for delegating medical record access

```
event medical_record_access(
      address requester,
      address owner,
      bytes conditions
      );
```

```
string[] patients;
string[] doctors;
mapping(doctors => patients) assignments;
int[] certificates;
mapping(doctors => certificates) proof-of-capabilities;
```

Fig. 6.16 A simple outline for recoding attributes

any further restrictions. There is no way for Fritz to ensure that only doctors assigned to Alice receive access. Unlike in [5], for our proposal, the delegation rules specified by Fritz can be used to restrict access as desired. Also, the proposal [5] requires centralized checking of the delegation on access where ours does not. A similar situation would apply if Alice's care in the clinic was passed from Fritz to another doctor. Our system would store attributes in the following format depicted in Fig. 6.16. This allows doctors to store official documents and at the same time permit the system to request them when necessary. In this case, may be multiple attribute providers are required that can provide attributes from different sources for validating identity of an entity. However, we leave this for a future work.

Note that the current implementation of this chapter is based on Solidity and the Ethereum blockchain, however the framework is generalized and can be implemented on any blockchain technology that supports smart contract execution. The implementation can easily be adapted to various use-case scenarios, e.g. that we discussed above. The programming code does not need to change to a higher extent to support a delegation. That said, the delegators and the delegatees can be any entities and they all rely upon certain attributes.

Note, in our system, security lies within the use of a blockchain network. This gives the system immutability of transactions, verifiable data, smart contracts and allows the system to be deployed in a decentralized environment. The environment removes the single point of failure found in similar centralized systems and makes the system overall more robust. Immutability and verifiable data ensure auditable and transparent transactions and attribute information stored within the system. They also give resource owners insurance that only verified users are given access. Additionally, the use of a blockchain gives the system access to smart contracts. These contracts give all users, including the resource owners, access to every request made through this system. Delegators, both the initial recipient of access and later delegates, may use smart contracts in the blockchain to specify conditions on further delegation. The use of smart contracts and requiring that all access requests be made through a root contract connects all legitimate contracts. This links all delegators and delegates. This allows for conditions to be compiled on top of each other as a request is made further down the chain meaning that all previous conditions must also be adhered to. Recall, in Sect. 6.2, we outline how the security properties of a blockchain (e.g. the immutable and verifiable data) are used to implement a delegation capability that are secure, verifiable and unforgeable. One crucial element that we use is ownership. The blockchain transactions and events tell us who generated and owns these events which, due to immutability, allows us to trust the context of their creation (when were they created in logical time and what are their purposes). Therefore, like regular capabilities, events give us the ideal solution to record distributed delegations and accesses.

One potential issue with our system is that it only guarantees pseudo anonymity. Attribute providers store attribute information on the blockchain which is linked to the user's Ethereum account address (wallet). This means that sensitive attributes pose the risk of privacy breach when stored on the system because a combination of these attributes may be used to de-identify a particular user and break the pseudo anonymity provided by the crypto-currency account address. This is not simply limited to attribute information. Every other curious user within the network will also know that a certain address has access to a certain resource. This imposes some limitations to our system, however, in this chapter we focus more on achieving the identity-less, asynchronous and decentralized properties and leave this issue for another future work.

6.8 Summary

IoT systems generate massive amounts of data, many of them may be sensitive (e.g. health related information), that comes from numerous devices, applications and services. This demands reliable connectivity and network scalability. Among others, protecting such data from unauthorized users and services requires proper security. In the previous chapters, we have noted that a significant security challenge for the IoT is the need for appropriate access control solution that are designed to meet the characteristics of these systems. To develop a secure access control model for the IoT, the propagation of access right delegation is a major issue. We have noticed that there are several issues that restrict a secure adaptation of access right delegation in IoT systems. For instance, the resource constrained nature of the IoT devices (e.g. limited memory capacity, battery power and computational processing capability) makes it difficult where the traditional access control mechanisms cannot be applied directly. The need for a delegation model leveraging the characteristics of an IoT system and at the same time to provide resiliency and flexibility in access right delegation are highly important. Furthermore, we observed that there is a need for the delegation of access control rights in IoT systems to be scalable, controlled in a fine-grained manner and recognize the inherently decentralized nature of such systems. Towards this, in this chapter, we used blockchain technology leveraging the distributed nature of access rights delegation for the large-scale IoT systems.

We have noted that the current models of delegations are based on the concrete identities of the users and the resources. In Chap. 5, we noted that the use of concrete identity for an entity in an IoT system is not an ideal condition. Previous approaches to delegation in the context of the IoT are mostly static in nature and do not consider the distributed essence of IoT systems. Moreover, they consider a highly resourceful environment for delegation. Most of the existing works that use RBAC and ABAC are highly centralized and can either grant or transfer operations (i.e. access rights) based on user-role or role-permission basis. When users and their various associations grow in a system, for example a large-scale IoT system, it is not an ideal choice to implement delegation mechanisms that depend solely upon either RABC or ABAC. While CapBAC addresses the issues of centralization by providing a decentralized nature of interactions, it is infeasible to define policies for each individual user. We observed that the highly centralized systems have the advantage of implicitly handling the capability transfer with simplicity and transparency, however once again, they are not dynamic in nature when distributing the delegation rights. We highlighted the use of attributes for authentication and authorization of entities in our system and did not consider the concrete identity of an entity. The attributes are used in smart contracts that are deployed on a blockchain, which helps in transforming the policy evaluation process to a fully distributed smart contract execution. Furthermore, in our model, a delegator can assign access conditions over the delegation process to a delegatee.

To investigate the significant features in our delegation e.g. identity-less, asynchronous and decentralized nature of communication, the examination of the integration of blockchain technology with IoT was the major scope of this chapter. We have seen that blockchain technology offers a secure and safe way to record and track a list of transactions for large numbers of devices in a highly transparent, auditable and efficient way by maintaining a P2P network. This is beneficial when overcoming the limitations of a centralized system for storing information. The use of blockchain technology also provides adequate support for data security in a distributed way which is suitable to employ with the IoT systems.

In the next chapter (Chap. 7), we intend to investigative the notion of trust in access control for the IoT. This will further help to enhance the delegation issues that we discussed in this chapter with the presence of uncertainty in the interactions between different entities in an IoT system.

References

1. A. Ouaddah, H. Mousannif, A. Abou Elkalam, A. Ait Ouahman, Access control in the internet of things: big challenges and new opportunities. Comput. Netw. **112**, 237–262 (2017). http://dx.doi.org/10.1016/j.comnet.2016.11.007
2. Q. Wang, N. Li, H. Chen, On the security of delegation in access control systems, in *Computer Security - ESORICS 2008*, ed. by S. Jajodia, J. Lopez (Springer, Berlin, 2008), pp. 317–332
3. S. Wohlgemuth, G. Müller, Privacy with delegation of rights by identity management, in *Emerging Trends in Information and Communication Security*, ed. by G. Müller. Lecture Notes in Computer Science, vol. 3995 (Springer, Berlin, 2006), pp. 175–190. http://dx.doi.org/10.1007/11766155_13
4. S. Gusmeroli, S. Piccione, D. Rotondi, A capability-based security approach to manage access control in the internet of things. Math. Comput. Model. **58**(5–6), 1189–1205 (2013). http://dx.doi.org/10.1016/j.mcm.2013.02.006
5. K. Hasebe, M. Mabuchi, A. Matsushita, Capability-based delegation model in RBAC, in *Proceedings of the 15th ACM Symposium on Access Control Models and Technologies*, ser. SACMAT '10 (ACM, New York, NY, USA, 2010), pp. 109–118. http://dx.doi.org/10.1145/1809842.1809861

6. O. Salman, I. Elhajj, A. Kayssi, A. Chehab, An architecture for the internet of things with decentralized data and centralized control, in *2015 IEEE/ACS 12th International Conference of Computer Systems and Applications (AICCSA)* (2015), pp. 1–8
7. J.L. Hernández-Ramos, A.J. Jara, L. Marín, A.F. Skarmeta Gómez, DCapBAC: embedding authorization logic into smart things through ECC optimizations. Int. J. Comput. Math. **93**(2), 345–366 (2016). http://dx.doi.org/10.1080/00207160.2014.915316
8. T. Kanter, S. Forsström, V. Kardeby, J. Walters, U. Jennehag, P. Österberg, Mediasense–an internet of things platform for scalable and decentralized context sharing and control, in *Proceedings of 7th IARIA International Conference on Digital Telecommunications (ICDT)* (Citeseer, 2012)
9. S. Nakamoto, Bitcoin: a peer-to-peer electronic cash system (2008)
10. R. Xu, Y. Chen, E. Blasch, G. Chen, A Federated Capability-based Access Control Mechanism for Internet of Things (IoTs) (2018). arXiv:1805.00825
11. J. Hernández-Ramos, J. Bernabe, M. Moreno, A. Skarmeta, Preserving smart objects privacy through anonymous and accountable access control for a M2M-enabled internet of things. Sensors **15**(7), 15 611–15 639 (2015). http://dx.doi.org/10.3390/s150715611
12. T.I. Kiviat, Beyond bitcoin: issues in regulating blockchain tranactions. Duke LJ **65**, 569 (2015)
13. X. Wang, X. Zha, W. Ni, R.P. Liu, Y.J. Guo, X. Niu, K. Zheng, Survey on blockchain for internet of things. Comput. Commun.**136**, 10–29 (2019), http://www.sciencedirect.com/science/article/pii/S0140366418306881
14. B. Awerbuch, D. Holmer, C. Nita-Rotaru, H. Rubens, An on-demand secure routing protocol resilient to byzantine failures, in *Proceedings of the 1st ACM Workshop on Wireless Security*, ser. WiSE '02 (ACM, New York, NY, USA, 2002), pp. 21–30. http://doi.acm.org/10.1145/570681.570684
15. J. Al-Jaroodi, N. Mohamed, Blockchain in industries: a survey. IEEE Access **7**, 36 500–36 515 (2019)
16. M. Maroufi, R. Abdolee, B.M. Tazehkand, On the convergence of blockchain and internet of things (iot) technologies (2019). arXiv:1904.01936
17. T.M. Fernndez-Carams, P. Fraga-Lamas, A review on the use of blockchain for the internet of things. IEEE Access **6**, 32 979–33 001 (2018)
18. M.A. Khan, K. Salah, Iot security: review, blockchain solutions, and open challenges. Future Gen. Comput. Syst. **82**, 395–411 (2018), http://www.sciencedirect.com/science/article/pii/S0167739X17315765
19. O. Novo, Blockchain meets iot: an architecture for scalable access management in iot. IEEE Int. Things J. **5**(2), 1184–1195 (2018)
20. R. Xu, Y. Chen, E. Blasch, G. Chen, Blendcac: a blockchain-enabled decentralized capability-based access control for iots (2018). arXiv:1804.09267
21. R. Xu, Y. Chen, E. Blasch, G. Chen, An exploration of blockchain enabled decentralized capability based access control strategy for space situation awareness (2018). arXiv:1810.01291
22. A. Manzoor, M. Liyanage, A. Braeken, S.S. Kanhere, M. Ylianttila, Blockchain based proxy re-encryption scheme for secure iot data sharing (2018). arXiv:1811.02276
23. R. Almadhoun, M. Kadadha, M. Alhemeiri, M. Alshehhi, K. Salah, A user authentication scheme of iot devices using blockchain-enabled fog nodes, in *2018 IEEE/ACS 15th International Conference on Computer Systems and Applications (AICCSA)* (IEEE, 2018), pp. 1–8
24. Y. Zhang, S. Kasahara, Y. Shen, X. Jiang, J. Wan, Smart contract-based access control for the internet of things. IEEE Int. Things J. 1–1 (2019)
25. A. Ouaddah, A.A. Elkalam, A.A. Ouahman, Towards a novel privacy-preserving access control model based on blockchain technology in iot, in *Europe and MENA Cooperation Advances in Information and Communication Technologies*, ed. by Á. Rocha, M. Serrhini, C. Felgueiras (Springer International Publishing, Cham, 2017), pp. 523–533
26. N. Tapas, G. Merlino, F. Longo, Blockchain-based iot-cloud authorization and delegation, in *2018 IEEE International Conference on Smart Computing (SMARTCOMP)* (2018), pp. 411–416
27. D. Di Francesco Maesa, P. Mori, L. Ricci, Blockchain based access control, in *Distributed Applications and Interoperable Systems*, ed. by L.Y. Chen, H.P. Reiser (Springer International Publishing, Cham, 2017), pp. 206–220

28. G. Ali, N. Ahmad, Y. Cao, M. Asif, H. Cruickshank, Q.E. Ali, Blockchain based permission delegation and access control in internet of things (baci). Comput. & Sec. (2019), http://www.sciencedirect.com/science/article/pii/S0167404819301208
29. M. Nuss, A. Puchta, M. Kunz, Towards blockchain-based identity and access management for internet of things in enterprises, in *Trust, Privacy and Security in Digital Business*, ed. by S. Furnell, H. Mouratidis, G. Pernul (Springer International Publishing, Cham, 2018), pp. 167–181
30. H. Shafagh, L. Burkhalter, A. Hithnawi, S. Duquennoy, Towards blockchain-based auditable storage and sharing of iot data, in *Proceedings of the 2017 on Cloud Computing Security Workshop*, ser. CCSW '17 (ACM, New York, NY, USA, 2017), pp. 45–50. http://doi.acm.org/10.1145/3140649.3140656
31. T. Le, M.W. Mutka, Capchain: a privacy preserving access control framework based on blockchain for pervasive environments, in *2018 IEEE International Conference on Smart Computing (SMARTCOMP)* (2018), pp. 57–64
32. J. Wainer, A. Kumar, A fine-grained, controllable, user-to-user delegation method in rbac, in *Proceedings of the tenth ACM symposium on Access control models and technologies* (ACM, 2005), pp. 59–66. https://doi.org/10.1145/1063979.1063991
33. Solidity, https://github.com/ethereum/solidity. Accessed 30 March 2017

Chapter 7
Integrating Trust to IoT Access Control

Abstract This chapter presents a trust management framework for the IoT that can improve access control mechanisms easing the decision-making process under uncertainty. The aim of this chapter to examine and introduce a well-defined trust model for IoT supporting attribute-based identity without the need for unique concrete identification of an entity. In the IoT, *things* often interact with one another often in unknown and uncertain circumstances. Therefore, in such systems, it is important to include mechanisms that can help in such interactions by overcoming this uncertainty. We argue that the notion of trust can assist in addressing such issues. Trust mechanisms allow entities to decide whether or not to interact with other entities. However, implementing trust in an IoT system is challenging due to the nature of the IoT systems themselves. It can be noted that one of the significant prerequisites of access control mechanisms is the notion of trust.

7.1 Introduction

In the IoT, *things* are expected to interact with one another, often autonomously. However, this poses some fundamental challenges for the security of IoT systems as the communication between the *things* often cannot be predicted in advance. In many cases, the interactions happen without the *things* that are communicating with one another directly possessing extensive information about each other. This information is based on the individual behaviour of the device, service or application. It can be supplemented by information provided by other *things* that have interacted with the entity in question. In access control decisions 'trust degree' plays an important role in both whether access is requested and whether it is granted [1].

The trust degree can be calculated in various ways. This may be based on a user's trust, device's trust or a system's trust. It may depend upon the context and the requirements of the system. The trust degree can be considered as a computational value that quantifies the relationship between the *trustor* (i.e. who is trusting) and the *trustee* (i.e. who is being trusted) within a specific context [2]. This further highlights the contextual information affecting in a communication. Trust management is a significant issue for an IoT user to be confident when accessing an IoT service.

S. Pal, *Internet of Things and Access Control*, Smart Sensors, Measurement and Instrumentation 37, https://doi.org/10.1007/978-3-030-64998-2_7

7.1.1 Problem Description

In Chap. 2 we discussed a range of issues that affect the development of trust in an IoT system. The heterogeneity and scale of the number of IoT devices, services and applications mean that previously unknown entities may often be encountered. Further, the nature of communications and other dynamic characteristics, make the IoT vulnerable from a security point of view. The interaction between the entities may be for a very short period of time and the entities may be interacting only once for their entire life-time. In other cases, the interactions may be very frequent and may be for a longer period of time. Further, IoT devices are in general resource constrained. They have limited battery power, memory capacity, processing speed, etc. [3]. Therefore, it is often not possible to store extensive interaction history or employ traditional heavy-weight security mechanisms in those devices for trust evaluation. Despite this there is still a need to measure how and when to trust an entity in the IoT. It could be argued that, given the potentially highly unpredictable nature of interactions in the general IoT, the need for trust mechanisms is even greater in such systems. However, in many cases, employing a centralized trusted authority is not feasible in the IoT [4].

The scale of the IoT systems make it necessary to question certain aspects of traditional approaches to trust management. Typically, trust management addresses the opinion that one entity (e.g. user, device, resource, *thing*) has of another singular entity. Given the scale of the IoT and the potential for sparse interaction patterns between individual *things*, holding trust information on a one-to-one basis appears both unsustainable and ineffective. It may be unsustainable in terms of the sheer amount of trust information and values that need to be maintained and it may be ineffective in that *things* may often encounter other *things* about which they have no information. Contextual information must be also taken into consideration at the time of trust value verification and update, further complicating the situation. Previously, in this book, we have argued that basing policy decisions and identity handling on attributes, rather than singular concrete identity, is a potential method for dealing with the nature of the IoT. This can also apply to trust management. In summary, rather than assessing trust on an individual *thing*, trust can be assessed on a set of attributes, and then applied to *things* that possess those attributes.

7.1.2 Contributions

In this chapter, we present a novel subjective logic based trust model for the IoT. In the model, the IoT *things* explicitly represent and manage ignorance as uncertainty in their trust relationship with other *things*. The major contributions of this chapter can be summarized as follows:

- We take uncertainties in an IoT system into consideration and propose a dynamic trust model for the IoT systems.

- We show that the proposed model is able to determine whether an IoT entity can be trusted for a specific service.
- We employ attributes to represent a set of services. In our model, an entity is validated based on attributes, rather than depending upon their concrete identities.
- We provide a detailed discussion of the trust management process and present different use cases to examine of the proposed model in real-world IoT scenarios.

7.2 Background

There are several proposals that have discussed the use of trust in an IoT context [5–7]. The *trust degree* (also referred as to the *trust value*) can be captured in various ways and based on the trust value computation, different design choices can be made [8, 9].

For instance, Bernabe et al. [10] propose a flexible trust-aware access control system for the IoT (TACIoT). This model is intended to serve as an end-to-end and reliable security mechanism for the edge IoT devices. In this model, trust values are calculated based on reputation, quality of service, security considerations and devices' social relationships, and calculated using fuzzy logic. The proposed model is composed of three main components, namely smart objects, authorization manager and trust manager. Smart objects are the smart IoT devices (e.g. sensors, actuators, etc.) that maintain social relationships composing different kinds of areas (e.g. personal, office, community, etc.). These smart objects can act as both CoAP client and server offering services (e.g. temperature, location, etc.) in an IoT environment. Every area has its own authorization manager which is responsible for creating authorization credentials (in terms of tokens) for the related smart objects. The PDP regulates the policy decision (based on XACML policies) for the authorization manager, and a 'Token Manager' generates the authorization credentials according to the authorization decisions made by the PDP. Every smart object is connected to a 'Trust Manager' which is responsible for assessing the trustworthiness degree of an entity. Given the resource-constrained nature of these smart objects, the 'Trust Manager' is treated as a separate network element, for example, a more powerful smart mobile phone. The 'Trust Manager' calculates trusts for both the service requester and the target. For the former, it determines the most trusted target for a specific service that is offered by many targets. For the latter, it determines the trust of a requester from the previous transactions. Finally, based on the trust value and the authorization credentials, an access decision is made. A detailed implementation is provided to support the proof of concept prototype. However, trust in this approach is tracked and calculated at the individual device level, impacting on the design's scalability.

Mahalle et al. [11] present a fuzzy approach to Trust Based Access Control (F-TBAC) with the notion of trust levels for identity management. The fuzzy approach for trust calculations is presented with the linguistic information of devices to address access control in the loT. A simulation based study is presented to demonstrate the proposed approach. Similarly to [10], in this model, trust is calculated based on experience, knowledge and recommendations. Finally, the proposed F-TBAC framework calculates the trust score and the trust score is then mapped to permissions to achieve an access control to a resource.

Saied et al. [12] propose a trust management framework for the IoT. The authors design a context-aware and multi-service trust management system that uses past experiences of interactions when calculating the trust value. The proposed solution takes into consideration the various resource capabilities of the *things* in a heterogeneous IoT architecture in order to establish a community of trusted elements that respect the objectives of the operation of a set of collaborative services. The trust management system is controlled and governed by a trust manager. The overall trust management and access control decisions follow five steps. First, the trust manager gathers information about the trustworthiness of the other nodes located in close distance for a service requesting node. Second, the trust management system issues recommendations about nodes to a service requesting node that intends to set up a collaborative service. Third, the service requesting node make a request for recommendations from the nearby nodes. Fourth, the service requesting node assesses the quality of each individual recommendation collected from the nearby nodes and a decision is made. Finally, the trust management system updates the information. A simulation based study is conducted to show the performance of the proposed system in managing trust and enforcing collaboration between the nodes.

Wang et al. [13] present a distributed trust management mechanism for the IoT. The authors extract three basic elements namely, service (i.e. defines the role of the trust management system), decision-making (i.e. making a decision to deliver a service) and self-organizing (i.e. selecting a decision by the trust management system), of trust management and then, based on a service model, a trust management framework is established for the IoT. Unlike the previous approaches, e.g. [10–12], in this framework, trust is employed based on the three core layers in an IoT architecture. These layers are the sensor layer (composed of smart IoT sensors and actuators), core layer (composed of networking and communication technologies) and application layer (composed of users' applications). For instance, trust in the sensor layer can be defined as a vector, as follows: $TS = \{X_s, TS_X, TS_{tar}, TS_{con}\}$. Where, X_s denotes the set of nodes $\{X_1, X_2, \ldots, X_N\}$, TS_X denotes a specific context, TS_{tar} denotes a specific target and TS_{con} denotes control information e.g. energy consumption or perception efficiency, etc. Similarly, trust can be calculated for core and application layers. Like [10, 11], a fuzzy approach is employed for trust value calculation.

Bao et al. [14–16] present several works on trust management for the IoT. In [14] a distributed and dynamic trust management protocol for IoT systems is presented. This deals with dynamic behaviour of nodes and detects misbehaving (i.e. both malicious and socially uncooperative) nodes status changes dynamically. The aim of this protocol is to dynamically adjust trust design parameter settings when the environment changes, therefore it is adaptive to changing environment conditions. This is measured after a trust assessment is done by each node, and the nodes maintain their own trust assessment for the other nodes. The proposed trust management protocol is encounter and activity based, i.e. after an encounter or interaction the trust value is updated. Two nodes in an encounter or interaction can directly observe each other and update their trust assessment accordingly. They can also recommend trust evaluation results to other nodes. Similarly to concepts in [10, 12], in this case, each node maintains various trust properties e.g. honesty, social cooperativeness and community interest. A formal model and simulation based experimental result are provided with a detailed attacks evaluation of the protocol.

In [15], authors extend the proposed work of [14] and show its adaptation in application to service composition in IoT context. In this approach, each device evaluates trust for a limited set of devices that holds the same interest using both direct observations and indirect recommendations. In [16], the authors extend the work presented in [15] in particular focused on community of interest (CoI) based social IoT systems where nodes can dynamically join and leave the system at any time. Where a group of nodes form a community based on their own interests. The major contributions of this proposal are the dynamic adaptation and scalability management of the trust management protocol in the IoT context. Dynamic adaptability property is demonstrated by showing that a new node in the community can quickly build its trust relationship with other nodes with desirable accuracy and convergence behaviour. To demonstrate the scalability, an efficient storage management strategy is introduced keeping the view of limited storage space of IoT nodes. In a storage space, each node can keep the corresponding trust information towards a subset of nodes according to their interest and storage space. A similar concept of trust management system is presented by Chen et al. in [17].

Sharma et al. [18] present a generic framework to manage trust for IoT systems considering qualitative and quantitative parameters. Similarly to concepts in [12], in this framework, the trust management encompasses multiple phases that are dedicated to perform different activities. Unlike [12], which consists of five phases, this framework contains four phases. They are information gathering, trust computation, trust dissemination and update and maintenance. No implementation is given to support the framework. Moreover, no contextual information is taken into consideration for trust computation.

There are several other works that discuss the modeling and design choice for how to incorporate trust in an IoT environment. For instance, Gago et al. [6] propose a theoretical approach for modelling of trust dynamics framework for IoT. The authors emphasize on the privacy and identity requirements for establishing trust in the IoT. No implementation is given to support the model. The trust evaluation is not dependent upon the feedback between the phases to ensure verification and traceability. Ferraris et al. [19], enhances the model presented in [6] and discuss a trust by design framework for an IoT entity considering the feedback from all the phases of the life-cycle of an IoT system. However, this approach is limited only to the theoretical framework and no details of the different phases and the transversal activities of the framework is given.

Trust and reputation in information and communication technologies have been considered as one of the prime factors for successful communication between two entities. For instance, Leister and Schulz [20] discuss a model for a trust indicator for the IoT. The authors use the concept of a priori and posteriori trust to give an indicator of how much a user in an IoT system can trust or distrust information that is provided by an IoT *thing*. Based on the trust indicator the user can decide the potential interactions to the IoT *things*. However, the model is highly centralized when evaluating the trust value indicator which is not an ideal base for IoT systems.

Wang et al. [21], discuss a trust model for access control in IoT that employs traditional ABAC systems that helps to consider the dynamic attributes of the entities in an IoT system. This model is composed of three main parts, namely, authentication, trust evaluation and access decision modules. The authentication module uses an identity-based authentication mechanism for restricting the illegal users in the system. The trust evaluation module uses specific trust threshold levels (e.g. good, normal and malicious) that are compared against the users' calculated trust values. Finally, based on the comparison of the cumulative trust value, an authorization is given by the access decision module. Note, this model is flexible in access control policy implementation that depends upon the traditional ABAC system. However, this model does not focus on how to build a trust model considering the trust relationship of two IoT entities for a given property under a specific set of contexts.

In summary, most of the trust management proposals discussed above are not an ideal base in the context of an IoT system given the resource constrained nature, dynamic characteristic in interactions and the presence of uncertainty when interacting with one another in an unknown situation. Unlike the aforementioned approaches, in this chapter, we consider various characteristics of an IoT system and use subjective logic for trust value evaluation. Recall, the employment of subjective logic is beneficial as it explicitly take uncertainty into consideration.

7.3 Context of the Model

Before going to the detailed discussion of our proposed trust management model, we first provide the characteristics of the model (cf. Sect. 7.3.1) and then present the basics of various subjective logic operators (cf. Sect. 7.3.2).

7.3.1 Characteristics

- Recall, in an IoT context, an entity may not know the identity of an interacting entity in advance. Therefore, in our trust model, we do not define a trusted entity by its unique concrete identity. We base our trust in an entity on a set of attributes rather than the entity's unique identity (cf. Chap. 5). Note, a device or a set of devices can be defined by one or a set of attributes. As discussed in the previous chapters of this book, this is beneficial given the dynamic characteristics of an IoT system.
- We assume that there are device managers who manage the devices. One device can be managed by multiple device managers or multiple devices can be connected to a single device manager.

- The way managers manage the devices can lead to a hierarchy. There may be a chain of device managers who are managing the devices.
- To handle uncertainty in interactions to an IoT context, we employ subjective logic for our trust model.

7.3.2 *Preliminaries*

As noted above, the proposed trust model is constructed using subjective logic. Recall, subjective logic is a probabilistic logic that explicitly takes uncertainty and belief ownership into account. It can be seen as a belief-reasoning calculus that is ideal for modeling and analysing situations involving uncertainty and relatively incomplete knowledge (cf. Chap. 2). Fundamentally, subjective logic uses special belief functions known as *opinion* and an opinion metric is given by $w = (b, d, u)$ where b, d and u denote belief, disbelief and uncertainty respectively. The corresponding values of $b, d, u \in [0, 1]$ and $b + d + u = 1$. The usefulness of subjective logic is the application of logic operators that allow the combination of different opinions into a combined single opinion.

Next, we briefly outline some basic subjective logic operation sets in the context of our trust model. We follow the notations used in [22].

Definition 7.1 (*Evidence to Opinion Mapping*) Let us consider an entity A requesting a service from another entity B. In this case, A would like to determine whether B satisfies a set of propositions x. Based on the past experiences of how B has satisfied x, A constitutes its opinion of B. If previously x was satisfied by B, then A marks this as a positive experience (pos). If x is not satisfied by B, then A marks this a negative experience (neg). Finally, if A is not able to determine whether B satisfies x or not, then A marks this an uncertain experience (unc). Note, as the experiences are individually marked as an evidence, the sum of the total pos, neg and unc would be equal to the total number of transactions between A and B. In such, A would then form its opinion using Eq. 7.1 below. ${}^{A}b_{B}^{x}$ represents A's belief on B about x, ${}^{A}d_{B}^{x}$ represents A's disbelief on B about x and ${}^{A}u_{B}^{x}$ represents as the A's ignorance on B about x.

$$
{}^{A}\omega_{B}^{x} = \{{}^{A}b_{B}^{x}, {}^{A}d_{B}^{x}, {}^{A}u_{B}^{x}\} \tag{7.1}
$$

$$
\begin{aligned}
{}^{A}b_{B}^{x} &= {}^{A}pos_{B}^{x} / \{{}^{A}pos_{B}^{x}, {}^{A}neg_{B}^{x}, {}^{A}unc_{B}^{x}\} \\
{}^{A}d_{B}^{x} &= {}^{A}neg_{B}^{x} / \{{}^{A}pos_{B}^{x}, {}^{A}neg_{B}^{x}, {}^{A}unc_{B}^{x}\} \\
{}^{A}u_{B}^{x} &= {}^{A}unc_{B}^{x} / \{{}^{A}pos_{B}^{x}, {}^{A}neg_{B}^{x}, {}^{A}unc_{B}^{x}\}
\end{aligned}
$$

Definition 7.2 (*Conjunction of Opinions*) The conjunction operator allows the merging of two opinions about a proposition into a single new opinion. It can be seen as the '*AND*' operation and is represented by the $\odot$ notation. Now consider, A has two opinions ${}^{A}\omega_B^x$ and ${}^{A}\omega_B^y$ of the entity B, for example, in two different contexts x and y. The conjunction ${}^{A}\omega_B^{x,y}$ of the opinions ${}^{A}\omega_B^x$ and ${}^{A}\omega_B^y$ represents A's opinion of B across both x and y and can be represented by Eq. 7.2. ${}^{A}b_B^{x,y}$ denotes A's belief in B in both x and y. Similarly, ${}^{A}d_B^{x,y}$ shows A's disbelief in B in both x and y and ${}^{A}u_B^{x,y}$ shows A's uncertainty about B in both x and y.

$$ {}^{A}\omega_B^{x,y} = {}^{A}\omega_B^x \odot {}^{A}\omega_B^y \tag{7.2} $$

This can also be written as:

$$
\begin{aligned}
{}^{A}b_B^{x,y}, {}^{A}d_B^{x,y}, {}^{A}u_B^{x,y} &= {}^{A}\omega_B^x \odot {}^{A}\omega_B^y \;\; where, \\
{}^{A}b_B^{x,y} &= {}^{A}b_B^x\, {}^{A}b_B^y \\
{}^{A}d_B^{x,y} &= {}^{A}d_B^x + {}^{A}d_B^y - {}^{A}d_B^x\, {}^{A}d_B^y \\
{}^{A}u_B^{x,y} &= {}^{A}b_B^x\, {}^{A}u_B^y + {}^{A}d_B^x\, {}^{A}u_B^y + {}^{A}u_B^x\, {}^{A}u_B^y
\end{aligned}
$$

Note that, if A generates an opinion for x at two different intervals of time (e.g. t_1 and t_2), then the conjunction of A's opinions for x at those times intervals represents its opinion at an imaginary time interval that represents both the time intervals t_1 and t_2.

Definition 7.3 (*Consensus of Opinions*) Consensus of opinions is defined as follows: *"the consensus rule for combining independent opinions consists of combining two or more independent opinions about the same proposition into a single opinion"* [23]. It is represented by the $\oplus$ notation.

Let us consider, A forms an opinion ${}^{A}\omega_B^x$ on B for certain proposition x and C forms another opinion ${}^{C}\omega_B^x$ on B for the same proposition x. Then, the consensus of these two opinions is given by Eq. 7.3.

$$ {}^{A,C}\omega_B^x = {}^{A}\omega_B^x \oplus {}^{C}\omega_B^x \tag{7.3} $$

This can also be written as:

$$
\begin{aligned}
{}^{A,C}b_B^x, {}^{A,C}d_B^x, {}^{A,C}u_B^x &= {}^{A}\omega_B^x \oplus {}^{A}\omega_B^y \;\; where, \\
{}^{A,C}b_B^x &= ({}^{A}b_B^x\, {}^{C}u_B^x + {}^{C}b_B^x\, {}^{A}u_B^x)/\xi \\
{}^{A,C}d_B^x &= ({}^{A}d_B^x\, {}^{C}u_B^x + {}^{C}d_B^x\, {}^{A}u_B^x)/\xi \\
{}^{A,C}u_B^x &= ({}^{A}u_B^x\, {}^{C}u_B^x)/\xi
\end{aligned}
$$

Here, $\xi = ({}^{A}u_B^x + {}^{C}u_B^x - {}^{A}u_B^x\, {}^{C}u_B^x)$ such that, $\xi \neq 0$. If $({}^{A}u_B^x, {}^{C}u_B^x) \to 0$, then ${}^{A,C}\omega_B^x$ can be denoted as follows:

$$^{A,C}b^x_B = (\xi\, ^Ab^x_B + {}^Cb^x_B)/\xi + 1$$
$$^{A,C}d^x_B = (\xi\, ^Ad^x_B + {}^Cd^x_B)/\xi + 1$$
$$^{A,C}u^x_B = 0 \ \ where,$$
$$\xi = {}^Au^x_B / {}^Cu^x_B$$

Definition 7.4 (*Discounting of Opinions*) Let us consider the following situation, where A has an opinion for B and B has an opinion for C with a proposition x. Then, A's opinion for C can be obtained by discounting B's opinion for C with the A's opinion for B. If $^A\omega^x_B = (^Ab^x_B, {}^Ad^x_B, {}^Au^x_B)$ and $^B\omega^x_C = (^Bb^x_C, {}^Bd^x_C, {}^Bu^x_C)$, then $^A\omega^x_C$ represents the discounted opinion of $^A\omega^x_B$ and $^B\omega^x_C$, as denoted in Eq. 7.4. It is represented by the $\otimes$ notation.

$$^A\omega^x_C = {}^Aw^x_B \otimes {}^Bw^x_C \tag{7.4}$$

This can also be written as:

$$^Ab^x_C, {}^Ad^x_C, {}^Au^x_C = {}^Aw^x_B \otimes {}^Bw^x_C \ \ where,$$
$$^Ab^x_C = (^Ab^x_B\, {}^Bb^x_C)$$
$$^Ad^x_C = (^Ab^x_B\, {}^Bd^x_C)$$
$$^Au^x_C = (^Ad^x_B + {}^Au^x_B + {}^Ab^x_B\, {}^Bu^x_C)$$

Definition 7.5 (*Disjunction of Opinions*) If A has an opinion for B and B has an opinion for C. Then assume, A's opinion for C can be obtained by disjunction of B's opinion for C with the A's opinion for B. If $^A\omega_B = (^Ab_B, {}^Ad_B, {}^Au_B)$ and $^B\omega_C = (^Bb_C, {}^Bd_C, {}^Bu_C)$, then $^Aw_{B\vee C}$ represents the disjuncted opinion of $^A\omega_B$ and $^B\omega_C$, as denoted in Eq. 7.5.

This is equivalent to the logical 'OR' operation and it is represented by the $\vee$ notation. Note, this operation is commutative and associative.

$$^A\omega_{B\vee C} = {}^A\omega_B \vee {}^A\omega_C \tag{7.5}$$

This can also be written as:

$$(^Ab_{B\vee C}, {}^Ad_{B\vee C}, {}^Au_{B\vee C}) = {}^A\omega_B \vee {}^A\omega_C \ \ where,$$
$$^Ab_{B\vee C} = {}^Ab_B + {}^Ab_C - {}^Ab_B\, {}^Ab_C$$
$$^Ad_{B\vee C} = {}^Ad_B\, {}^Ad_C$$
$$^Au_{B\vee C} = {}^Ad_B\, {}^Au_C + {}^Au_B\, {}^Ad_C + {}^Au_B\, {}^Au_C$$

Definition 7.6 (*Negation of Opinions*) The negation of opinions is defined as follows: *"a negation of an opinion about a proposition consists of inverting the belief* *and disbelief components while keeping the ignorance component unchanged"* [23]. Let us consider that ${}^{A}\omega_x = (b_x, d_x, u_x)$ is A's opinion for an entity B about a proposition x (when x is true). Then the negation of opinion can be defined as Eq. 7.6. It can be seen as the equivalent to the logical unary operation 'NOT' and it is represented by the $\neg$ notation.

$$ {}^{A}\omega_B^{\neg x} = ({}^{A}b_B^{\neg x}, {}^{A}d_B^{\neg x}, {}^{A}u_B^{\neg x}) \tag{7.6} $$

$$ where,\ {}^{A}b_B^{\neg x} = {}^{A}d_B^{x},\ {}^{A}d_B^{\neg x} = {}^{A}b_B^{x},\ {}^{A}u_B^{\neg x} = {}^{A}u_B^{x} $$

7.4 Proposed Trust Model

In this section, we present our proposed Trust Model (TM) and discuss its formalization. The trust model TM can simply be defined as $TM = (E, TR, OP)$. Where, E denotes the set of entities that share one or more trust relationships, TR represents the set of trust relationships between the entities under a certain set of conditions and OP denotes the set of operations that manages the TR. As noted earlier, there are many ways to define trust, however, in the context of our model, we follow the proposal discussed in [22]. Next, we present a formal definition for each component of the model.

7.4.1 Entities

Definition 7.7 (*Entities*) Consists of different members of the model that share one or many trust relationships with one another.

In our model, we employ two types of entities. First, who is trusting (trustor) and second what is being trusted (trustee). These are all *things* within an IoT system. The trustor is the entity that is attempting to determine the level of trust. Amongst other possible examples it might be a user device attempting to determine whether to request a service or it might be a service-providing device attempting to determine whether to trust a requestor. The trustee is the target of the trust enquiry. Trust evaluation can be carried out before making a decision as to whether to make or accede to an access request.

7.4.2 Trust Relationship

Definition 7.8 (*Trust Relationship*) It is defined by the following tuple,

$$TR = (A, B, Rel, M, pos, neg, unc, T, Cnt, Cat, \Theta) \tag{7.7}$$

The tuple TR represents that an entity (A) trusts another entity (B) in a certain trust relationship (Rel) with the total number of opinions with positive (pos), negative (neg), uncertain (unc) experiences calculated by an evidence mapping function (M) with certain trust type (T) in a given context (Cnt) and category (Cat) at a given time (Θ).

- A is an individual entity. In our model, it is an entity that is determining its trust in another entity.
- B is the entity that is being trusted by the entity A. Note, in our model, B is represented by a set of $\{attribute, value\}$ pairs. Multiple actual *things* may conform to the attribute set B and be trusted at the same level. The trusting entity A must decide what sets to keep and to record trust information against.
- Rel is the relationship between the trustor and the trustee. For example, one or a set of devices that are managed by one or a set of different device managers. In other words, the relationship between a device D_1 and a device manager DM_1, where the D_1 is managed by the DM_1. This can be used to distinguish particular sets of entities with the same attribute sets.
- M is the evidence mapping operation for a certain trust relationship TR. The mapping operator takes the collected evidence (pos, neg and unc experiences) and forms an opinion as per equation 1 above.
- pos is the total number of positive experiences associated with a certain trust relationship TR.
- neg is the total number of negative experiences associated with a certain trust relationship TR.
- unc is the total number of uncertain experiences associated with a certain trust relationship TR.
- T is the trust type. For example, direct trust, recommended trust or derived trust.
- Cnt represents the context of the trust evaluation. It is defined by a specific instance in which trust is evaluated. Similar to other uses of context, in this book, it can be seen as the set of environmental attributes. For example, a certain time and location or both of them, or certain devices that are present in a specific location.
- Cat is the specific category of trust. For example, the level of satisfaction in an interaction or certain certification that is trusted by the entity A.
- Θ is the set of time stamps at which the different trust types (e.g. direct, recommended and derived) and trust experience values (e.g. pos, neg and unc) were last updated for a certain trust relationship TR.

7.4.3 System Operation

In this section, we explain how our proposed trust model would function. Imagine an IoT system that is composed of a number of user devices, devices and device managers. All of these are *things*. Devices and device managers exist within a particular domain. Devices must be connected to at least one device manager. However, a device can be connected to multiple device managers. Each device manager may be connected to one or more devices.

When a device (user device or other *thing*) wants to access a device (*thing*), it needs to communicate with the device manager of that resource. First it determines its trust level in the device, or, more precisely, in a set of attributes the values of which conform to those held by the device. As multiple devices may all hold the same attribute values this avoids the need of maintaining individual trust values for each device. For example, the user device may record trust against the set (class = 'light', location = 'E6A building'). That is, all the lights in the E6A building. As noted, the user device does then not need to individually track its trust in each such light. It could, if it prefers, track trust in a more fine-grained manner, for example, adding another attribute to the set (e.g. 'floor') and individually tracking trust in the group of lights in each floor. One advantage of this approach is that as lights are added to a location (e.g. floor) the user device automatically has a level of trust in that new device. This may be useful in other situations, where, for example, a device manager is managing a set of sensors located in an outdoors environment, some of which are located in a dry area and some in a swamp. Assume that the user device is able to distinguish between these two locations by the value of the device's 'location' attribute and that devices in the swamp are typically unreliable due to degradation caused by the environment. The user device could maintain two trust values, one for devices in the swamp and one for devices in the dry area. New devices added would be trusted according to their 'location'.

The device manager also evaluates its overall trust value in the requesting device in order to grant or deny the requested service. As with the requesting device, the device manager does not need to maintain individual trust values in the user device, but instead can base its trust on the attributes possessed by the user device.

The attributes possessed by a device (user device, device manager, device) may be any physical or functional properties of an entity (e.g. location, date, time, or all of them together). This allows, for example, the requesting user device to assess its trust in the device based on its previous experience of devices with those attributes and similarly for the device manager's trust in the requesting device. This will simplify trust management by reducing the number of separate records that need to be maintained.

In our approach, an important aspect is attribute-based identification of devices. That is, we identify devices in terms of the attributes they satisfy. Hence, within a domain a device manager DM_1 manages a collection of devices that satisfy a given set of attributes. In other words, DM_1 manages devices which are characterized by a set of attributes. We can now say, $DM_1 = an\ entity$ and $D_1 = \{sets\ of\ attributes\}$.

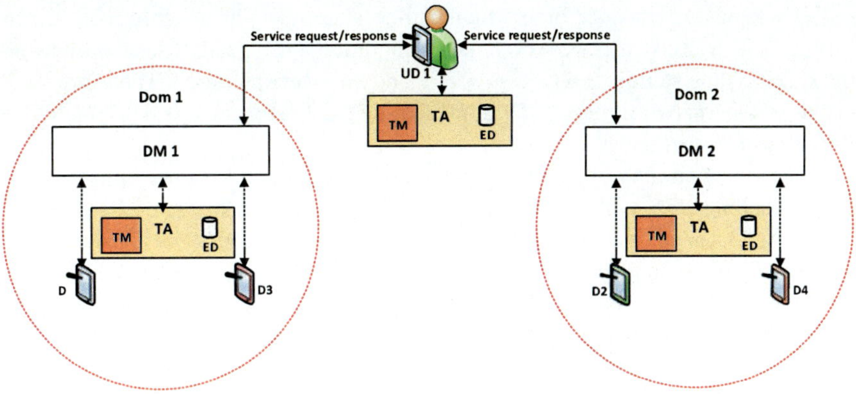

Fig. 7.1 A simple system model for trust management

Let us now consider the situation more formally. For simplicity, assume there are two domains (Dom_1 and Dom_2), Dom_1 is managed by the device manager DM_1 and Dom_2 is managed by the device manager DM_2. Device D_1 belongs to Dom_1 and therefore controlled by the DM_1. Similarly, device D_2 belongs to Dom_2 and controlled by the DM_2. Assume further that D_1 and D_2, through their device managers, are offering similar services and possess a set of attributes in common. In Fig 7.1, we illustrate the system model. Note, each user device and domain manager maintains their own trusted authority (labeled as TA). Each TA consists of a trust manager (labeled as TM) and an evidence database (labeled as ED). TM is responsible for overall trust value calculation and ED is responsible for storing the collected evidences.

Assume a user device UD_1 requests services from D_1 through D_1's manager DM_1. The trustor (i.e. UD_1 in this case) decides what attributes are relevant for a certain evidence mapping operation (M) and therefore what set of attributes are required. Call this set of $\{attribute, value\}$ pairs AV_1. The evidence mapping operation on a certain TR denotes the opinion of one entity on another entity. In our case, it represents UD_1's opinion on D_1 based on the total number of positive (pos), negative (neg) and uncertain (unc) experiences on the set of attributes it holds to which D_1 conforms, that is AV_1. UD_1 wants to see whether there are past interactions with this attributes set AV_1 (more properly, with devices which conform to AV_1). UD_1 collects the total number of opinions e.g. pos,neg and unc on AV_1 for the evaluation mapping function (M). This gives the direct trust.

The specific value returned by the evidence mapping function (M) is dependent upon the context (cnt) and category (cat) of the TR between UD_1 and D_1. For instance, the values of opinions (pos, neg, unc) can be varied based on context x and context y. UD_1 may also gather recommendations from other devices for the attributes value set AV_1. It can then calculate the derived trust from the direct and recommended trusts.

UD_1 needs to consider the particular time stamp (Θ) at which direct, recommended and derived trusts were evaluated. In addition, UD_1 needs to take into account the specific time stamps at which pos, neg and unc were recorded. Therefore, UD_1 evaluates the trustworthiness of D_1 by evaluating the TR for AV_1. DM_1 can calculate its trust in UD_1 similarly.

Over time a number of successful interactions occur and trust builds up between the interactors. UD_1 will associate its trust with D_1 with the attributes D_1 possesses - AV_1. At a later point UD_1 requests services from D_2 through D_2's manager DM_2. Assume D_2 also has the attributes and values specified by AV_1. As D_2 has the same attributes as D_1, UD_1's direct trust in D_1 is extended to D_2. Note that the user device UD_1 selects which attribute sets to hold and measure trust in. The user device can be as fine or coarse-grained in this decision as it wishes. This means that it can ensure that its trust is not applied too broadly but only to entities that are sufficiently similar and related to those with which it has previously interacted. Note that trust experience is recorded based on individual contexts, which also assists in managing trust relationships.

DM_2 may wish to evaluate trust in UD_1 in determining whether to allow access to the service provided by D_2. It may base this, in part on recommendations from DM_1 and from its own interactions with other devices with a set of attributes held by UD_1. Note that the latter case would be direct trust and, again, the trusting device is in control of which attributes a device must hold to be considered under that collection of trust evidence.

7.4.4 Summary of Trust Types

Our proposal employs three types of trust, namely, direct trust, recommended trust and derived trust, as discussed in Sect. 7.5. In this section, we provide a summary outline of the way that these trust types function in our system based on the aforementioned system operation.

- *Summary of Direct Trust:*

(a) In our framework, a device is characterized by its attributes, therefore, UD_1's trust on D_1 is characterized based on the attributes of D_1.
(b) UD_1's trust on D_2 is same as UD_1's trust on D_1 as long as both the D_1 and D_2 have the same set of $\{attribute, value\}$ pairs and the context, category and relationship value are the same.
(c) UD_1's trust on D_2 can be different from that of D_1 if any of context, category and relationship is different, even if both D_1 and D_2 have the same set of attributes.

- *Summary of Recommended Trust:*

(a) If UD_1 asks for a recommendation for D_1 from UD_2 this will be based on a set of $\{attribute, value\}$ pairs. If UD_2 has interacted with D_1 then it will have experiences recorded against the $\{attribute, value\}$ set and can make a recommendation.

(b) If UD_1 asks for a recommendation for D_1 from UD_2 this will be based on a set of $\{attribute, value\}$ pairs. If UD_2 has not interacted with D_1 but has interacted with one or more other devices that conform to that $\{attribute, value\}$ set then it will have an experience with that set of attributes and can make a recommendation.

(c) If UD_2 has not interacted with D_1 or with any other device possessing the relevant $\{attribute, value\}$ then it cannot make a recommendation.

- *Summary of Derived Trust:*

(a) In all cases, derived trust is calculated from direct and recommended trust as described below in Sect. 7.5. Note, in the absence of direct trust the derived trust is equal to the recommended trust. Similarly, in the absence of recommended trust the derived trust is equal to the direct trust.

7.5 Trust Evaluation

In this section, we discuss the trust evaluation process for our proposed trust model. In our model, we use the representation of trust as an opinion metric. The opinion metric is denoted by $\omega = (b, d, u)$, where b denotes belief, d denotes disbelief and u denotes uncertainty for a given trust relationship TR.

Definition 7.9 (*Evidence Mapping*) The evidence mapping operator M for a certain trust relationship TR represents the opinion of one entity about another entity. In our case, it can be seen as the opinion of one entity about a set of attributes. The opinion is collected in the form of positive (pos), negative (neg) and uncertain (unc) experiences of the collected evidence based on the previous interactions. Recall, the evidence mapping operation is given in Eq. 7.1.

Definition 7.10 (*Opinion Decay*) It is a function that represents the dynamic nature of trust. Over a given time frame, the value of trust for a certain entity varies based on certain factors e.g. context, category, relationship, etc. The opinion decay helps to represent the trust as time progresses. Note that the decay function $(\Psi_{k,\Delta})$ is used to calculate new opinion (ω_{new}) after decay from an old opinion (ω_{old}). This can be represented as in Eq. 7.8.

$$\omega_{new} = \Psi_{k,\Delta}[\omega_{old}] \tag{7.8}$$

where, k represents the decay rate and $0 < k \leq 1$. Δ represents the time difference at which trust is evaluated and the opinion was last updated for a certain entity.

$\Psi_{k,\Delta}$ can be represented as:

$$b_{new} = b_{old}[1 - e^{-(k.\Delta)}]$$
$$d_{new} = d_{old}[1 - e^{-(k.\Delta)}]$$
$$u_{new} = u_{old} + [(b_{old} + d_{old}) - (b_{new} + d_{new})]$$

$\Psi_{k,\Delta}$ should be chosen so that ω_{new} does not decay too rapidly. The value of $\Psi_{k,\Delta}$ may need to be capped to avoid extreme decay in the event of long periods between updates.

Definition 7.11 (*Total Opinion*) The total opinion of an entity A about a set of attributes B for a given context cnt_i, category cat_j and relationship rel_k is given by the combined evidence collected for that context, category and relationship.

Now let us assume, $({}^{A}pos_{B(cnt_i,cat_j,rel_k)}, {}^{A}neg_{B(cnt_i,cat_j,rel_k)}, {}^{A}unc_{B(cnt_i,cat_j,rel_k)})$ represents the evidence associated with the trust relationship TR of an entity A for a set of attributes B for the given context cnt_i, category cat_j and relationship rel_k. Based on the collected evidence, the evidence mapping function M is used to calculate the opinion of the TR at time θ. This can be represented as in Eq. 7.9.

$$ {}^{A}\omega^{\theta}_{B(cnt_i,cat_j,rel_k)} = \{{}^{A}b^{\theta}_{B(cnt_i,cat_j,rel_k)}, {}^{A}d^{\theta}_{B(cnt_i,cat_j,rel_k)}, {}^{A}u^{\theta}_{B(cnt_i,cat_j,rel_k)}\} \tag{7.9}$$

This can be regarded as the opinion calculated over all evidence collected. That is evidence collected between time θ and system start time 0 (zero). Opinion can also be calculated over evidence collected in a specified time period, e.g. between time θ and time $\theta - t$. This can be written as in Eq. 7.10.

$$ {}^{A}\omega^{\theta,\theta-t}_{B(cnt_i,cat_j,rel_k)} = \{{}^{A}b^{\theta,\theta-t}_{B(cnt_i,cat_j,rel_k)}, {}^{A}d^{\theta,\theta-t}_{B(cnt_i,cat_j,rel_k)}, {}^{A}u^{\theta,\theta-t}_{B(cnt_i,cat_j,rel_k)}\} \tag{7.10}$$

Equation 7.9 could then be re-written with θ replaced by θ, 0, but for convenience we will retain the formation as in Eq. 7.9.

Definition 7.12 (*Direct Trust*) Direct trust is the belief that an entity A holds on a set of attributes B for a given context, category and relationship based on its past experiences with B. The direct trust in a certain context cnt_i, category cat_j and relationship rel_k is calculated based on the opinion held at the time of service request θ and the total opinion prior to service request at time $(\theta - t)$. This can be represented as in Eq. 7.11.

$$ {}^{A-dir}\omega_{B(cnt_i,cat_j,rel_k)} = {}^{A}\omega^{\theta,\theta-t}_{B(cnt_i,cat_j,rel_k)} \odot \Psi_{k,t}[{}^{A}\omega^{\theta-t}_{B(cnt_i,cat_j,rel_k)}] \tag{7.11}$$

where, ${}^{A}\omega^{\theta,\theta-t}_{B(cnt_i,cat_j,rel_k)}$ and ${}^{A}\omega^{\theta-t}_{B(cnt_i,cat_j,rel_k)}$ are, respectively the opinion calculated at time θ based on evidence collected since the last update, i.e. $(\theta - t)$, and the opinion calculated at the time of the last update $(\theta - t)$, based on all evidence collected up to that time. The latest opinion at time θ is combined with the all previous experiences before the time θ, where, θ, $(\theta - t) \in \Theta$. The opinion at time $(\theta - t)$ is decayed to emphasize more recent experiences.

Definition 7.13 (*Recommended Trust*) It is the belief that an entity A holds on a set of attributes B for a certain context, category and relationship based on the recommendations obtained from its peer entities' past experiences. The notation ${}^{A-rec}\omega_{B(cnt_i,cat_j,rel_k)}$ represents the overall recommended opinion of an entity A on B computed from the individual opinions of A's recommenders for the particular context, category and relationship. This can be represented as in Eq. 7.12.

$$ {}^{A-rec}\omega_{B(cnt_i,cat_j,rel_k)} = (F_{R_1} \otimes \Psi_{k,t_1}[{}^{R_1}\omega^{\theta-t_1}_{B(cnt_i,cat_j,rel_k)}]) \oplus \cdots \oplus (F_{R_m} \otimes \Psi_{k,tm}[{}^{Rm}\omega^{\theta-tm}_{B(cnt_i,cat_j,rel_k)}]) \tag{7.12} $$

Note, a decay function is applied to every recommended opinion. For each decayed opinion of a recommender, a critical factor F is introduced to the corresponding recommendations. It signifies how much the entity A values each recommendation. The F of $F_{R_1} \ldots F_{R_m}$ are attached to the decayed opinions of recommenders $R_1 \ldots R_m$ respectively using a discounting operator. Note that the value of t used for the decay of each recommendation is different as each recommender may have updated its opinion at a different point in the past.

Definition 7.14 (*Derived Trust*) It is the belief an entity A builds on a set of attributes B for a given context, category and atomic trust relationships e.g. direct and recommended. The derived opinion for a given context cnt_i, category cat_j and the relationship rel_k is the combination of direct and recommended opinions for these context, category and relationship. This can be represented as in Eq. 7.13. For example, a user device wishing to obtain a service from a $thing$ may obtain a recommendation from that $thing$'s manager. The relationship value here will be different between the users device and the manager due to their different relationships with the device. The user device can include the manager's recommendation as just outlined, with an appropriate weighting value.

$$ {}^{A-der}\omega_{B(cnt_i,cat_j,rel_k)} = {}^{A-dir}\omega_{B(cnt_i,cat_j,rel_k)} \oplus {}^{A-rec}\omega_{B(cnt_i,cat_j,rel_k)} \tag{7.13} $$

Note that the derived trust is the combination of both the direct and recommended trust. Therefore, it can be obtained by the combination of Eqs. 7.11 and 7.12 together. Importantly, as we discussed above, in the absence of a direct trust, the derived trust is equal to the recommended trust. Similarly, in the absence of a recommended trust, derived trust is equal to the direct trust.

Definition 7.15 (*Total Derived Trust*) It is the belief an entity A builds on a set of attributes B for all contexts, categories and relationships.

In calculating the derived trust in another entity the trustor may not wish to limit itself to a particular, context, category or relationship. This may be, for example, in the case where the amount of evidence is limited. Instead, the trustor may use wider information. For example, the total derived trust over all contexts for a particular category and relationship can be represented as in Eq. 7.14.

$$^{A-der}\omega_{B(cat_j,rel_k)} = ^{A-der}\omega_{B(cnt_1,cat_j,rel_k)} \oplus ^{A-der}\omega_{B(cnt_2,cat_j,rel_k)} \oplus \cdots \oplus ^{A-der}\omega_{B(cnt_I,cat_j,rel_k)} \tag{7.14}$$

where, I is the total number of contexts. The total derived trust can also be written as:

$$^{A-der}\omega_B = \oplus(^{A-der}\omega_{B(cnt_i,cat_j,rel_k)}), \forall 1 \leq i \leq I, \forall 1 \leq j \leq J, \forall 1 \leq k \leq K \tag{7.15}$$

This idea can also be used in the calculation of either or both of direct and recommended trust. For example, trustors may not wish to limit the recommendations they receive to one set of values for context, category and relationships. The approach taken in Eqs. 7.14 and 7.15 for forming a consensus of opinions across a range of contexts, categories and relationships can then be used to assemble a variety of recommendations.

7.6 Trust Comparison

The major intention of using trust in the IoT context is to reduce the uncertainty in the communication and secure authorization of the resources. The trust model presented in this chapter is based on opinion metrics. The comparison of opinion metrics is used in order to make an authorization to a certain service. For instance, the derived trust of a platform should be higher than an expected opinion threshold by which a request will be serviced. In Sect. 7.7, we provide practical examples. Next, we represent the comparison operator as follows:

Definition 7.16 (*Comparison of Opinions*) An opinion comparison operator $\geq_{ope}$ for two given opinions ope_1 and ope_2, where $ope_1 \geq_{ope} ope_2$, holds if $b_1 > b_2$, $d_1 < d_2$ and $u_1 < u_2$. In such case, we can say that a trust value for ope_1 is higher than the threshold trust value presented by the ope_2.

When an entity A receives a request from another entity B for a certain service, A may wish to determine whether B should be provided with that service. To determine this, A computes its direct trust for the B using Eq. 7.11 for the relevant values of context, category and relationships. A also gathers recommendations for B for the same context, category and relationship and possibly for other contexts, categories and relationships as well. Note, each recommended opinion is decayed for the time elapsed from when the last recommendation was recorded. The decay time Δ may vary for each recommender. It can also be possible that the different decay rate k is used for each recommender by A. The recommended trust is calculated using Eq. 7.12. The derived trust then can be computed combing the direct and the recommended trust using Eq. 7.13, possibly extended as described under total derived trust. For every service that is provided, A assigns authorization policies which includes a threshold value ope_{th}. This can be seen as an opinion constant for a given context, category and trust relationship. Using the comparison operator $\geq_{ope}$, A now compares the derived opinion with the threshold opinion. If the derived opinion is greater than the threshold opinion, A trusts B and the request will be serviced. Note, in the reverse case, B deciding whether it should trust A for the provision of a service, the calculation proceeds in the same manner.

7.7 Use Case Scenarios

In this section, we discuss two example scenarios to show the potential where our trust evaluation framework would function.

7.7.1 Scenario 1

In Chap. 1, we noted that implantable IoT-enabled medical devices can be a potential vector for attackers to breach a patient's private information or even damage the patient's health. For instance, attacks made on a pacemaker can cause cardiac arrest [24].

Suppose, in a smart home, a patient's health is periodically measured and communicated to a hospital where the patient is treated. Now consider, the pacemaker of a patient suddenly sends an urgent notification regarding the cardiac condition of the patient (e.g. a cardiac arrest). This information arrives at the central response serve) that is managed and maintained by the hospital. We assume that this server includes a trust management module that now evaluates the overall trust between the pacemaker and the server. This will help to determine the seriousness of the situation and corresponding actions to be taken. For this, the sensor module must be satisfied with a certain threshold of the calculated trust value before a decision is made (discussed in Sect. 7.6). The direct trust can be calculated based on the interactions between the pacemaker and the server. In case of the recommended trust, the server collects recommendation information from other IoT smart devices that have previous interaction with the pacemaker. In our proposed trust model, the server is the trustor and the pacemaker is the trustee. Now consider the following two cases.

Case 1: In this case, let us assume that there is no recommender available. In other words, the weight for the recommended trust is zero. Therefore, the derived trust is equals to the direct trust. We assume that the time of the decision making is March 20, 2019 at 18:00:00 hours and opinion at the sensor module was last updated 20 days prior to request. The following trust relationship is available:

$$(A, B, Rel, Cnt, Cat, March\ 20\ 2019, 18:00:00, [0.42, 0.32, 0.26], 8, 6, 5)$$

Therefore, the direct trust can be seen as follows:

$$\Psi_{1,20/365}.{}^{A}\omega^{\theta-20}_{B(cnt_i,cat_j,rel_k)} = (0.17, 0.55, 0.22)$$

Case 2: In this case, the hospital management module collects some recommendations from other devices that have previous interactions with the pacemaker. We consider two devices (i) smartphone of the doctor who installed the pacemaker (denoted as R_1) and (ii) an IoT-enabled smart device of the manufacturer of the pacemaker (denoted as R_2). In some other cases, it could also be possible that other IoT devices can provide recommendations about the particular pacemaker. For example, an ECG monitoring device situated in the patient's home that has previous interactions with the pacemaker. For simplicity, we only consider the above mentioned two devices in this case and that values of context, category and relationship all match. It is important to note that the trust calculation will make use of the most recent recommendations. Typically, the previous recommendations can be stored in a hospital database server. Therefore, in this case, the derived trust is calculated from the direct and the recommended trust. Let us assume that there are one direct interaction and two recommendations (one from smartphone and one from manufacturer's IoT smart device). The following trust relationships are available. We consider the direct trust as in case 1 discussed above. Opinions are decayed at the rate of 100% and the recommendations are collected in the past 6 and 10 days. The following trust relationships are available:

$$(R_1, B, Rel, Cnt, Cat, March\ 24\ 2019, 18:00:00, [0.26, 0.72, 0.00], 5, 14, 0)$$
$$(R_2, B, Rel, Cnt, Cat, March\ 20\ 2019, 18:00:00, [0.22, 0.67, 0.11], 4, 12, 2)$$

The recommended opinions and decayed recommended opinions of recommenders R_1 and R_2 based on previous experiences are:

$$^{R_1}\omega^{\theta-6/365}_{B(cnt_i,cat_j,rel_k)} = (0.26, 0.72, 0.00),\ \Psi_{1,6/365}.{}^{A}\omega^{\theta-6/365}_{B(cnt_i,cat_j,rel_k)} = (0.13, 0.36, 0.50)$$
$$^{R_2}\omega^{\theta-10/365}_{B(cnt_i,cat_j,rel_k)} = (0.22, 0.67, 0.11),\ \Psi_{1,10/365}.{}^{A}\omega^{\theta-10/365}_{B(cnt_i,cat_j,rel_k)} = (0.11, 0.32, 0.55)$$

Therefore, The total recommended trust from the R_1 and the R_2 is (0.16, 0.47, 0.36). Now, if we calculate the derived trust in this case, it is (0.19, 0.60, 0.16), where, $\Psi = 0.49$.

7.7.2 Scenario 2

In this scenario, we consider an indoor hospital environment. In a particular ward there are several patient monitoring devices installed. For example, IoT-enabled devices are attached to the patient's bed. These devices can periodically send information to the hospital database regarding a patient's physical and clinical information e.g. blood glucose level, heart rate, blood pressure, location within the ward, etc. Any malicious activities in one of these devices can cause potential threat to the patient's health. For instance, unauthorized control to a patient's insulin pump and alter the dose of the medicine or similar activities [24]. Doctors and nurses carry IoT-enabled portable devices that are able to communicate to these various IoT devices within the ward and the hospital database.

Now suppose, an insulin pump sends a critical signal e.g. low level of blood glucose of a patient admitted in a certain ward. In this case, the information is transmitted to the hospital central server from where the patient monitoring control is managed and maintained. Therefore, in this case, the insulin pump is the trustee and the hospital central server is the trustor. Before taking an action in this situation, the central server wants to ensure the seriousness of the situation. In such, it performs a trust evaluation on the insulin pump from where the request is received. Similar to the previous scenario (i.e. Scenario Sect. 7.7.1), there are two cases possible:

Case 1: In this case, there is no recommender is available, so the weight for the recommendation trust is zero. Hence, the derived trust is equals to the direct trust. We assume that the time of the decision making is June 10, 2019 at 18:00:00 hours and opinion at the central server was last updated 5 days prior to request. The following trust relationship is available:

$$(A, B, Rel, Cnt, Cat, June\ 10\ 2019, 18:00:00, [0.47, 0.37, 0.16], 9, 7, 3)$$

Therefore, the direct trust can be seen as follows:

$$\Psi_{1,5/365}.{}^{A}\omega^{\theta-5}_{B(cnt_i,cat_j,rel_k)} = (0.20, 0.59, 0.15)$$

Case 2: In this case, the central server collects some recommendations from other devices that are able to provide recommendations for that particular insulin pump. Assume five are available (i) an IoT-enabled smart device (denoted as R_3) of the doctor who is treating the patient, (ii) and (iii) two IoT-enabled smart devices (denoted as R_4 and R_5) of two nurses who controls the dose of the medicine in different shifts and (iv) and (v) two nearby patient monitoring devices (denoted as R_6 and R_7) to the insulin pump. Once again, we consider the direct trust as discussed in case 1 above. Opinions are decayed at the rate of 100% and the recommendations are collected in the past 2, 8 and 12 days. The following trust relationships are available:

$$(R_3, B, Rel, Cnt, Cat, June\ 10\ 2019, 18:00:00, [0.30, 0.60, 0.10], 6, 12, 0)$$
$$(R_4, B, Rel, Cnt, Cat, June\ 02\ 2019, 18:00:00, [0.31, 0.62, 0.08], 4, 8, 1)$$
$$(R_5, B, Rel, Cnt, Cat, June\ 02\ 2019, 18:00:00, [0.33, 0.56, 0.11], 3, 5, 1)$$
$$(R_6, B, Rel, Cnt, Cat, May\ 30\ 2019, 18:00:00, [0.45, 0.45, 0.09], 5, 5, 1)$$
$$(R_7, B, Rel, Cnt, Cat, May\ 30\ 2019, 18:00:00, [0.46, 0.38, 0.15], 6, 5, 2)$$

The recommended opinions and decayed recommended opinions of recommenders R_3 to R_7 based on previous experiences are:

$$^{R_3}\omega^{\theta-2/365}_{B(cnt_i,cat_j,rel_k)} = (0.30, 0.60, 0.10),\ \Psi_{1,2/365}.{}^{A}\omega^{\theta-2/365}_{B(cnt_i,cat_j,rel_k)} = (0.29, 0.58, 0.10)$$

$$^{R_4}\omega^{\theta-8/365}_{B(cnt_i,cat_j,rel_k)} = (0.31, 0.67, 0.11),\ \Psi_{1,8/365}.{}^{A}\omega^{\theta-8/365}_{B(cnt_i,cat_j,rel_k)} = (0.30, 0.60, 0.08)$$

$$^{R_5}\omega^{\theta-8/365}_{B(cnt_i,cat_j,rel_k)} = (0.33, 0.56, 0.11),\ \Psi_{1,8/365}.{}^{A}\omega^{\theta-8/365}_{B(cnt_i,cat_j,rel_k)} = (0.33, 0.54, 0.11)$$

$$^{R_6}\omega^{\theta-12/365}_{B(cnt_i,cat_j,rel_k)} = (0.45, 0.45, 0.09),\ \Psi_{1,12/365}.{}^{A}\omega^{\theta-12/365}_{B(cnt_i,cat_j,rel_k)} = (0.44, 0.44, 0.09)$$

$$^{R_7}\omega^{\theta-12/365}_{B(cnt_i,cat_j,rel_k)} = (0.46, 0.38, 0.15),\ \Psi_{1,12/365}.{}^{A}\omega^{\theta-12/365}_{B(cnt_i,cat_j,rel_k)} = (0.45, 0.37, 0.15)$$

Therefore, The total recommended trust collected from the R_3 to R_7 is (0.31, 0.28, 0.39). Now, if we calculate the derived trust in this case, it is (0.26, 0.56, 0.12), where, $\Psi = 0.48$.

We consider that a trustor will provide a service if the derived trust on the trustee is in the following threshold opinion of (0.60, 0.30, 0.10). In other words, the derived trust value is compared against the above mentioned threshold opinion. In the first scenario, we noted that the derived trust is (0.19, 0.60, 0.16) and in the second scenario the derived trust is (0.26, 0.56, 0.12). Therefore, in both cases, the trustor is not satisfied with given threshold opinion of the trustee.Therefore, the trustor will not provide a service to the trustee. This also highlights that the trustor is able to detect the falsified claims from the devices in the proposed trust model.

7.8 Summary

In this chapter, we have proposed a theoretical framework for a trust model for the IoT considering the highly dynamic nature and other distinct characteristics of such systems. We have noted that trust is an important factor for establishing communication between different entities in uncertain conditions. This very true in the case of the IoT, as the interacting entities may not know each other's identities in advance or may not encounter each other before. Trust can help address this lack of certainty. This may come from direct interactions between these entities or any other recommendation that comes from a trusted third party. We argued that in the IoT the *things* must have the ability to make decisions autonomously without any human intervention. This end trust in IoT has been recognized as an important issue for processing and handling of data.

The establishment of trust and to define a flexible trust management system for the IoT is in its infancy. In most of the cases, the interactions between the entities in an IoT system occur without enough knowledge of the interacting entities. The information of the entities in an application may not only come from its behaviors but also the device properties in a certain context. In addition, the trustworthiness of the different layers in an IoT architecture play a crucial part to the overall trust management due to the need for reliable communication among the layers. This reinforces the need for interoperability, robustness and dynamicity for the IoT trust modelling.

There are several trust management proposals that discussed the centralized trust management systems for the IoT. These models aim at collecting trust values from the entities and process them in a central system, based on the employed trust modelling framework (cf. Sect. 2.6 of Chap. 2). The centralized trust management mechanisms can provide a sustainable means towards the trust management in the IoT. However, the centralized approach to IoT trust management is not an ideal solution due to the characteristics of the IoT. Furthermore, the central systems are not always efficient enough to capture the trust values of the misbehaving nodes according to different contexts. To overcome the limitations of a centralized trust management system, dynamic trust management systems are proposed. In this approach, an individual IoT entity is capable of performing the trust evaluation at runtime and autonomously. A few other approaches consider the social aspects in trust management systems based on social relationships e.g. cooperativeness, honesty, community interests, etc. This paradigm is also known as the Social IoT (SIoT).

In our proposed trust model, an entity can make an authorization decision using different trust types, e.g. direct trust, recommended trust and derived trust. Significantly, we used derived trust as the default design of trust comparison computed with the direct and recommended trusts. Our model can efficiently adopt uncertainty present in an IoT system using subjective logic for trust value calculation. We argued that every trust management system should have some threshold that must be satisfied before a trustor offers a service to the trustee. Therefore, we demonstrated the practicability of our proposed model using real-world use case scenarios where trust valued are evaluated based a threshold.

In the next chapter (i.e. Chap. 8), we will present a summary of the research findings that are discussed in previous chapters of this book.

References

1. F. Almenárez, A. Marín, C. Campo, C. García R, Trustac: trust-based access control for pervasive devices, in *Security in Pervasive Computing*, ed. by D. Hutter, M. Ullmann (Springer, Berlin, 2005), pp. 225–238
2. N.B. Truong, U. Jayasinghe, T.-W. Um, G.M. Lee, A survey on trust computation in the internet of things. J. Korean Inst. Commun. Inf. Sci. (JKICS) **33**(2), 10–27 (2016)
3. V. Gligor, J.M. Wing, Towards a theory of trust in networks of humans and computers, in *Security Protocols XIX*, ed. by B. Christianson, B. Crispo, J. Malcolm, F. Stajano (Springer, Berlin, 2011), pp. 223–242
4. J. Sabater, C. Sierra, Review on computational trust and reputation models. Artif. Intell. Rev. **24**(1), 33–60 (2005). https://doi.org/10.1007/s10462-004-0041-5
5. I. Ud Din, M. Guizani, B. Kim, S. Hassan, M. Khurram Khan, Trust management techniques for the internet of things: a survey. IEEE Access **7**, 29 763–29 787 (2019)
6. C. Fernandez-Gago, F. Moyano, J. Lopez, Modelling trust dynamics in the internet of things. Inf. Sci. **396**, 72–82 (2017), http://www.sciencedirect.com/science/article/pii/S0020025517305364

7. A. Altaf, H. Abbas, F. Iqbal, A. Derhab, Trust models of internet of smart things: a survey, open issues and future directions. J. Netw. Comput. Appl. (2019), http://www.sciencedirect.com/science/article/pii/S1084804519300839
8. Z. Yan, P. Zhang, A.V. Vasilakos, A survey on trust management for internet of things. J. Netw. Comput. Appl. **42**, 120–134 (2014), http://www.sciencedirect.com/science/article/pii/S1084804514000575
9. L. Fritsch, A.-K. Groven, T. Schulz, On the internet of things, trust is relative, in *Constructing Ambient Intelligence*, ed. by R. Wichert, K. Van Laerhoven, J. Gelissen (Springer, Berlin, 2012), pp. 267–273
10. J. Bernal Bernabe, J.L. Hernandez Ramos, A.F. Skarmeta Gomez, Taciot: multidimensional trust-aware access control system for the internet of things. Soft Comput. **20**(5), 1763–1779 (2016). https://doi.org/10.1007/s00500-015-1705-6
11. P.N. Mahalle, P.A. Thakre, N.R. Prasad, R. Prasad, A fuzzy approach to trust based access control in internet of things. Wireless VITAE **2013**, 1–5 (2013)
12. Y.B. Saied, A. Olivereau, D. Zeghlache, M. Laurent, Trust management system design for the internet of things: a context-aware and multi-service approach. Comput. & Secur. **39**, 351–365 (2013), http://www.sciencedirect.com/science/article/pii/S0167404813001302
13. J.P. Wang, S. Bin, Y. Yu, X.X. Niu, Distributed trust management mechanism for the internet of things, in *Applied Mechanics and Materials*, vol. 347 (Trans Tech Publ, 2013), pp. 2463–2467
14. F. Bao, I.-R. Chen, Dynamic trust management for internet of things applications, in *Proceedings of the 2012 International Workshop on Self-aware Internet of Things*, ser. Self-IoT '12 (ACM, New York, NY, USA, 2012), pp. 1–6. http://doi.acm.org/10.1145/2378023.2378025
15. F. Bao, I.-R. Chen, Trust management for the internet of things and its application to service composition, in *IEEE International Symposium on a World of Wireless* (Mobile and Multimedia Networks (WoWMoM), 2012), pp. 1–6
16. F. Bao, I. Chen, J. Guo, Scalable, adaptive and survivable trust management for community of interest based internet of things systems, in *2013 IEEE Eleventh International Symposium on Autonomous Decentralized Systems (ISADS)* (2013), pp. 1–7
17. I. Chen, F. Bao, J. Guo, Trust-based service management for social internet of things systems. IEEE Trans. Dependable Secure Comput. **13**(6), 684–696 (2016)
18. A. Sharma, E.S. Pilli, A.P. Mazumdar, M.C. Govil, A framework to manage trust in internet of things, in *2016 International Conference on Emerging Trends in Communication Technologies (ETCT)* (2016), pp. 1–5
19. D. Ferraris, C. Fernandez-Gago, J. Lopez, A trust-by-design framework for the internet of things, in *2018 9th IFIP International Conference on New Technologies, Mobility and Security (NTMS)* (2018), pp. 1–4
20. W. Leister, T. Schulz, Ideas for a trust indicator in the internet of things, in *The First International Conference on Smart Systems, Devices and Technologies* (2012)
21. J. Wang, H. Wang, H. Zhang, N. Cao, Trust and attribute-based dynamic access control model for internet of things, in *2017 International Conference on Cyber-Enabled Distributed Computing and Knowledge Discovery (CyberC)* (2017), pp. 342–345
22. A. Nagarajan, V. Varadharajan, Dynamic trust enhanced security model for trusted platform based services. Future Gen. Comput. Syst. **27**(5), 564–573 (2011), http://www.sciencedirect.com/science/article/pii/S0167739X10002013
23. A. Jøsang, Artificial reasoning with subjective logic (1997)
24. WIRED, How the internet of things got hacked (2015), https://www.wired.com/2015/12/2015-the-year-the-internet-of-things-got-hacked/. Accessed 01 Oct 2017

Chapter 8
Conclusion and Future Work

Abstract This chapter summarizes the major findings of the research. In this chapter, we revisited the contributions and list a number of open research questions and provide insights for future research directions.

8.1 Conclusion

The emergence of the IoT has already produced significant changes in our everyday lives, where everything and anything can be connected and communicated in the cyber-physical world. With the proliferation of smart mobile devices, intelligent sensors, wearable devices, ubiquitous Internet and cloud computing, the use of the IoT is growing at an increasing rate. However, we noted that this growth poses numerous challenges for the designers and users of these systems. One significant challenge is the provision of security within the IoT. The high mobility of *things*, the potential scale of the systems in the number of *things* and users combined with dynamic network topology and wireless communication mediums create a challenging environment. This is only exacerbated by the limitations in device memory, battery-life and processing capacity, arguing against the use of 'heavy-weight' security architectures.

We found that the massive scale of the integration of heterogeneous devices and services in an IoT system means that none of the commonly used access control mechanisms (e.g. RBAC, ABAC, CapBAC, etc.), in isolation, can achieve efficient management of access control policies and enforcement of authorization decisions. Moreover, the intrinsic features of traditional access control approaches may be difficult to implement within the resource constrained IoT devices. There is a requirement that, whatever access control mechanism is employed, it should be usable as well as sufficient to protect the privacy, integrity and confidentiality of the system and its components. In an IoT system, information should easily be available and accessible to the legitimate users. However, the information should be protected, allowing only authorized users to control and manipulate the data.

In this research, we assert that there is a need for rethinking the requirements of an IoT security architecture that achieves fine-grained access control requiring

S. Pal, *Internet of Things and Access Control*, Smart Sensors, Measurement and Instrumentation 37, https://doi.org/10.1007/978-3-030-64998-2_8

minimum mechanisms for policy enforcement and their management, that enables secure access control for billions (and possibly trillion) of *things* which can access and be accessed in a heterogeneous environment. This in turn, will help to develop a scalable, dynamic and flexible access control architecture for the IoT.

8.2 Research Findings

This research presents a new and practical approach to the use of attributes for authentication and authorization of entities for an IoT system. It develops the concept of an access control architecture for large-scale IoT for flexible and fine-grained policy management. Major research findings as listed as follows:

In Chap. 2, we have discussed related work related to the state of the art access control mechanisms for the IoT. This included a wider view of access control issues and their potential integration to IoT and provided a detailed discussion of the IoT security requirements in a systematic way. We also briefly introduced the notion of identity, access control delegation and the significance of trust in an IoT context.

In Chap. 3, we raised an important question of how to build a secure access control architecture based on the identified access control requirements in the context of the IoT. To answer the question, we used a use-case of smart healthcare system where several actors are involved and proposed an access control architecture for IoT. We emphasized that fully centralized control over an IoT system is not an ideal base for providing QoS in many ways, including fault tolerance, distribution of access rights or event the support for heterogeneity. We have outlined the design of a general access control system for the IoT that combines elements of attributes, roles and capabilities to achieve streamlined policy management. We noted that traditional access control mechanisms which, in isolation, incapable of providing fine-grained and flexible access control for the IoT. The proposed design is partially decentralized, which helped to achieve better performance for IoT systems using light-weight security mechanisms. In our system, access rights are embodied in capabilities. The capabilities are provided to users on request, based on the attributes of users and the roles which those users give them membership of. Once a capability is obtained, the user attributes do not need to be checked again while the capability remains valid. Significantly, the edge devices need only to check the capability, avoiding any communication with a central system at that point, including any need for repeated attribute evaluations. We intended to see the use of light-weight security protocols (e.g. symmetric key based approach) and their potential applicability to the constrained IoT devices. At the same time, we compared the employment of asymmetric key based approach to the developed architecture.

In Chap. 4, we demonstrated the feasibility of the proposed access control architecture using a physical experimental setup. We used commodity hardware for building the constrained IoT devices. We employed ESP8266-12E microcontroller in terms of price, performance, connectivity and ease of use. With numerical results, we showed how our proposal significantly improved the performances leveraging the advantages of RBAC, ABAC and capabilities to a fine-grained level. We used RBAC to improve the relationship between the user and policies by using specific roles. ABAC improved the fine-grained access control by reducing the number of policies required for the identification process. Finally, the use of capabilities improved the proliferation of user's identity without requiring heavy-weight policy management. We conducted both symmetric and asymmetric key based experimentation and listed the differences. Indeed, the asymmetric key based approach of implementation delivers better security than a symmetric key based approach, but there is a trade-off between the resource constrained nature of IoT devices and the need for security provisioning. Our experimental results confirmed the suitability of the light-weight security protocol within the architecture. However, the symmetric key based design depends upon pre-registration of the user device and the things with the central management system. To avoid this would require the use of asymmetric key cryptography which would increase the round-trip time. Our results suggested that the proposed style of distributed authorization system for constrained IoT devices can be an alternative to fully centralized authorization systems for large-scale systems e.g. the IoT.

In Chap. 5, we examined the core concept of digital identity and its provisioning to IoT identity. We further enhanced the findings to analyze the requirements for an IoT identity model supporting attribute based identity. We answered the importance question of how to precisely identify the exact services to which many entities will seek access. In a large-scale system like the IoT, it is challenging due to the characteristics of the IoT systems, where it is difficult to predict, in advance, which entities will interact and require access to services from which entity. The technical question was how to build a formal model of IoT identity based on attributes, rather than depending upon a concrete identity of an entity. Another major concern was the presentation of minimum attributes for an identification while keeping confidential the other related information. To achieve this, we have investigated various representations of digital identity and its suitability when addressing identity in an IoT context. Given the analysis, this led us to build an identity model for the IoT where we have presented a novel idea of IoT identity from the *thing's* perspectives. The *thing* can, for example, be a device, an application, a human user or even an organization. We have further illustrated a formal model of IoT identity based on the different components of an identity management framework in a more systematic fashion. Our proposal was also supported by specific use-cases examples.

In Chap. 6, we investigated the significant issues of transferring access rights through delegation in an IoT context. This is important, because in large-scale and dynamic systems, e.g. the IoT, delegation is crucial in ensuring flexible, fine-grained and responsive access to resources by allowing users to propagate access in a controlled fashion. We noted that the previous approaches to delegation in the context of the IoT are mostly static in nature and do not consider the distributed essence of IoT systems. This led us to proposed an identity-less, asynchronous and decentralized delegation model for flexible and ease transfer of access rights in the IoT. To achieve this, we used blockchain technology. Our prime goal was to demonstrate how to facilitate managing and accessing IoT resources without the need for a central trusted authority. We proposed and analyzed a novel capability-based delegation model for large-scale IoT systems using blockchain. We showed that a blockchain platform provides all the necessary support to implement a delegation process that is identity-less, asynchronous and decentralized by nature. The basic idea was to use capabilities to propagate access rights and use blockchain for communication between various entities. The delegated capabilities were checked by the edge IoT *things* upon access. Once again, our design was based on attributes and does not depend upon a concrete identity of the IoT *things*, this provides significant flexibility when managing resources at scale. Furthermore, we demonstrated how our proposal took advantage of decentralized nature of blockchain networks and support the asynchronous nature in communication.

In Chap. 7, we proposed a trust model for the IoT. We used the dynamic notation of trust to examine access control in such IoT systems. We showed how to improve access control mechanisms easing the decision-making process under uncertainty. We used subjective logic for the proposed trust model and introduced a well-defined trust model for IoT supporting attribute-based identity. As to the main focus of this book, we did not use unique concrete identification of an entity. That said, our proposal was based on attributes where entities are represented by the attributes they hold. We employ various subjective logic operators to define the proposed model. We showed how trust can be calculated based on opinions in an uncertain situation. We demonstrated the usefulness of the proposed model using different real-world IoT scenarios. Our finding showed that it is possible to build such a trust model by considering the dynamic characteristics on a large-scale IoT system.

8.3 Open Research Questions and Future Work

In this research, we have discussed access control, identity management and secure access right delegation for the large-scale systems like the IoT. However, there are several open research questions which future research would need to address. This includes the needs for supporting heterogeneity in IoT access control and identity management to provide uniform and easy to manage solutions across the range of technologies that are deployed in the IoT. An IoT system must support the scale of the number of devices, users, applications and their associated services. This indicates the need for an efficient policy management that can achieve both in terms of the size of the policy database and ability to manage newly encountered users and devices. Policy mechanisms will need to be flexible and employ new levels of abstraction. There is always a trade-off between centralized and distributed architectures. There is need to balance the need for effective and secure security provisioning against the low resource provisioning of devices and likely user requirements. Further, a system should situate the security components as close as to the local IoT devices by developing a scalable and structured security infrastructure for the IoT. One significant feature of the IoT is the vast number of devices of which it consists. This brings advantages and disadvantages. A primary drawback is the resource limited nature of many of these devices, especially wearable ones which cannot be permanently linked to a mains power source. Battery life therefore becomes a concern. There is a need to reach a balance between light-weight solutions and adequate provision of IoT security given the resource-constrained nature of the IoT devices. In the future we are looking to build the prototype using a more powerful microcontroller. The revocation of capabilities is another piece of future work that needs to be investigated further in the context of our proposed access control architecture. In our proposed development of the blockchain based access control delegation we situate attribute providers in a public blockchain infrastructure. To provide adequate privacy for the attributes further investigation may be conducted. Finally, further developments of the proposed trust model, and detailed simulations based upon it, need to be investigated.